Advances in Anatomy, Embryology and Cell Biology
Ergebnisse der Anatomie und Entwicklungsgeschichte
Revues d'anatomie et de morphologie expérimentale

50 · 5

W0050637

Editors

A. Brodal, Oslo · W. Hild, Galveston · J. van Limborgh, Amsterdam
R. Ortmann, Köln · T. H. Schiebler, Würzburg · G. Töndury, Zürich · E. Wolff, Paris

Z. Halata

The Mechanoreceptors of the Mammalian Skin Ultrastructure and Morphological Classification

With 11 Figures

Springer-Verlag Berlin Heidelberg New York 1975

PD Dr. Zdeněk Halata
Anatomisches Institut der Universität Hamburg
2000 Hamburg 20, Martinistraße 52
Bundesrepublik Deutschland

Library of Congress Cataloging in Publication Data. Halata, Z. 1939—. The mechanoreceptors of the mammalian skin ultrastructure and morphological classifikation. (Advances in anatomy, embryology and cell biology; 50.5) Bibliography: p. Includes index. 1. Mechanoreceptors. 2. Mammals—Anatomy. I. Title. II. Series. [DNLM: 1. Mechanoreceptors—Anatomy and histology. 2. Skin—Innervation. W1 AD433K v. 50 fasc. 5 / WR101 H157 m]. QL801.E67 vol. 50, fasc. 5 [QL938.M4] 574.4'08s. [599'.01'858]. 74-34232.

ISBN-13: 978-3-540-07097-9   e-ISBN-13: 978-3-642-45465-3
DOI: 10.1007/978-3-642-45465-3

Contents

_____

To my Teacher Prof. Dr. med. Přemysl Poláček, DrSc.

5

Introduction

Whilst most of the senses (hearing, sight, smell and taste) have their own organs, the tactile sense is dependent on the sensory nerve endings of the peripheral processes of the nerve cells in the spinal ganglia. These nerve endings are distributed over the entire body. They vary in number and structure according to the nature of the tissue. For instance, the quantitative innervation of the mucosa differs from the innervation of the periosteum or the articular capsules. The skin and its related tissues are relatively richly innervated, but here too there are regional differences. Some areas, such as the skin of the back, have relatively few nerve endings, whilst other parts (e.g. the skin of the fingers) are richly innervated.

Most authors describe the nerve endings systematically from the surface of the epidermis to the lower layers of the dermis. On the basis of the topographical criteria, we differentiate between epidermal and dermal nerve endings.

Epidermal Nerve Endings

Authors in the 19th century (Langerhans, 1868; Podcopaev, 1869; Eimer, 1871; Merkel, 1875, 1880; Mojsisovics, 1876; Ranvier, 1880; Cybulsky, 1883; Hoggan, 1884; Retzius, 1892, 1894), and at the beginning of the 20th century (Huss, 1900; Leontowitsch, 1901; Botezat, 1902a, b, 1903, 1907, 1908, 1912; Tretjakoff, 1902, 1911; Pincus, 1905; Bielschowsky, 1907; Kadanoff, 1924a, b, 1928; Boeke, 1925, 1926, 1932, 1933a, 1934, 1940; Woolard, 1933) tended to concentrate on detailed descriptions of the courses of the sensory nerve fibers and their peripheral ramifications. It was recognised early on that in addition to the so-called "free" nerve endings in the epidermis, there were nerve endings showing a thickening of the nerve in contact with a specialised cell, a so-called "tactile cell" (Merkel, 1875, 1878, 1880; Bonnet, 1878; Ranvier, 1880a, b; Hoggan, 1884; Dogiel, 1891, 1903, 1905; Retzius, 1892; Szymonowicz, 1895, 1897, 1933; Ksjunin, 1899; Tretjakoff, 1902, 1911; Pincus, 1902, 1905; Botezat, 1908, 1912; Vincent, 1913; Kadanoff, 1924b, 1928; Boeke, 1925, 1933b, 1934). The frequently contradictory views found in these works are due to different methods of displaying the nerve endings. A critical evaluation is difficult nowadays as the findings were recorded exclusively by drawings.

Further advances in methods (and especially improved resolution of the microscope) in the course of the last twenty-five years have resulted in an improved knowledge of the sensory nerve endings. More recent authors have refrained from a detailed description of the peripheral ramifications of the nerve fibres as well as from classification of the vast number of sub-types of these nerve fibres and have concentrated instead on general questions concerning the relationship of the endings to their environment, their structure, and their topographical position. The "free" nerve endings have been described, with the aid of light microscopy, by: Weddel *et al.* (1954), Goethe (1956), Ono (1956), Miller, Ralston and Kasahara (1958, 1960), Cauna (1959), Lele and Weddel (1959),

Fitzgerald (1961), Cauna and Alberti (1961), Quilliam (1966), Sinclair (1967), Ridley (1969), Jabonero and Moya (1971), Niewöhner and van der Zypen (1972), and Loo and Kanagasunteram (1972). Electron microscopy has demonstrated that the "free" nerve endings occupy an extracellular position in the epidermis (Munger, 1965; Orfanos, 1965a, b, 1967; Hagen and Werner, 1966a, b, 1967; Hagen, 1967; Böck, 1971; Nafstad, 1971a, b; Kadanoff, 1971a, b, c; Tsuji, 1971; Halata, 1972a).

More attention has been paid to the Merkel endings. These consist of specialised "tactile cells" with thickened nerve fibres. They have been observed by means of light microscopy by Melaragno and Montagna (1953), Kawamura (1954), Cauna (1958, 1959), Straile (1960), Mann and Straile (1965), Quilliam and Armstrong (1963), Kawamura et al. (1964), and Quilliam (1966). They have been described with the aid of the electron microscope by Cauna (1962, 1966), Iggo and Muir (1963, 1969), McGavran (1964), Munger (1965, 1966, 1971), Smith (1965, 1967, 1970), Andres (1966a, b, 1969), Quilliam (1966), Halata (1970, 1971, 1972a), Breathnach and Robins (1970), Munger, Pubols and Pubols (1971), Breathnach (1971a, b), Hashimoto (1971, 1972), Nafstad (1971b, 1972), Nikai, Rose and Cattoni (1971), Kidd, Krawczyk and Wilgram (1971), Lyne and Hollis (1971). The physiological work of Iggo (1963, 1966), and Iggo and Muir (1969) has demonstrated that Merkel endings are slowly adapting mechanoreceptors of Type II a. The problems of their origin and the mechanism of transmission of impulses from Merkel cells to the nerve endings are as yet unsolved.

Dermal Nerve Endings

The structure of the dermal nerve endings is much more complicated than that of the epidermal nerve endings. In addition to the simple "free" dermal nerve endings, whose function has not yet been fully clarified (it is assumed that they respond to warm and cold), the dermis contains complicated nerve endings which are known as corpuscular receptors. According to Botezat (1912), the "free" nerve endings differ from the corpuscular ones in that they have no capsules. These so-called corpuscles differ in size and structure, which has misled several authors into classifying these endings. Krause (1881) differentiated between 13 types of sensory corpuscles. Botezat (1912) describes 6 types, with two or three variants for each. The best subdivision is that of Stöhr (1928), who differentiates three types of corpuscular end-organ. The contradictory nomenclature of the corpuscles is due to their variability. Two main types were distinguished on the basis of the arrangement of the nerve fibres in the corpuscles, and that of the auxiliary cells surrounding the nerve fibres. The first group comprises the bodies whose axons are in the form of spiral or coil. Depending on their situation, these endings are known either as Meissner's corpuscles (Wagner's or Wagner-Meissner corpuscles), or as a genital corpuscles. These endings have been studied by a number of authors during the last century and the beginning of this one (Wagner and Meissner, 1852; Kölliker, 1852; Meissner, 1853, 1855, 1895; Huxley, 1854; Tomsa, 1865; Fischer, 1876; Merkel, 1880; Ranvier, 1880b; Krause, 1891; Dogiel, 1892, 1903; Smirnow, 1893; Szymonowicz, 1895, 1933; Crevatin, 1901; Leontowitsch, 1901; Ruffini, 1902; van der Velde, 1909; Lefebure, 1910; Botezat, 1912; Heringa, 1920; Boeke, 1933b; Jalowy, 1936). Publications since 1950 include those of Cauna (1953, 1954, 1956a, b, 1958), and Weddel et al. (1954). The new electron microscope investigations have revealed that the structure of Meissner's corpuscle is the same as that of the genital corpuscles and of Dogiel's corpuscle (Cauna and Ross, 1960; Patrizi and Munger, 1965; Cauna, 1966; Poláček and Malinovský, 1971; Halata, 1972c).

The second group of these corpuscles invariably has an inner core containing the terminal of the axon. These endings were first discovered by Vater (1741), and rediscovered by Pacini (1835). They are referred to in the literature as Vater-Pacini corpuscles or, more usually, as Pacinian corpuscles. They are exceedingly variable, not only with regard to size but also to form, and in respect of the number of their inner cores. The literature on the subject is very extensive. The works of Cauna and Mannan (1958, 1959) contain a comprehensive survey of the older literature. Recently Malinovský (1966a, b, c, 1970) published light microscope studies on the variability of these endings in the glabrous skin of the cat. Electron microscopy of these endings in the mesentery of the cat has been carried out by Pease and Quilliam (1957), Poláček and Mazanec (1966), and their blood supply was studied by Pallie et al. (1970). Chouchkov (1971a, b) has described the degeneration after severance of the afferent nerves, and the regeneration. These endings are not confined to the mesentery of the cat but occur throughout the mammalian body, e.g. in the articular capsules (Poláček, 1961, 1965, 1965), in the wall of the urinary bladder (Shehata, 1970, 1972), in the skin of the fingers by man (Cauna, 1956) and even in the middle ear (Gussen, 1970). Similar endings, which differ from the typical Pacinian corpuscles by their much smaller size, have been found with the sinus hair of the cat (Andres, 1966a, b), in the nasolabial region (Ormea and Goglia, 1969; Poláček, 1969) and in the lips of the cat (Spassova, 1970). These endings have also been found in animals other than the cat—such the mouth skin of cattle (Walter, 1956, 1961, 1962; Walter and Hebel, 1966; Hebel and Schweiger, 1967), the nose of the mole (Quilliam, 1966; Halata, 1972b), and in the paws of the tree-shrew (Andres, 1969).

Their development has been described by Poláček and Halata (1970). Their findings show that immature endings are generally situated in the nerve fascicles. During the maturation of the corpuscle (which in the cat takes up to three months), the Schwann cell develops round the axon terminal without the formation of a myelin sheath and so forms an inner core. The capsule of the corpuscle develops from the perineurium of the nerve fascicle. The studies of Poláček (1968) have shown that these endings differ in size according to the depth of their position. The deeper the endings, the larger they are.

Physiological classification includes this group among the rapidly adapting receptors. According to Schmidt (1971), the larger Pacinian corpuscles are acceleration detectors. They receive vibration impulses (Quilliam and Armstrong, 1963; Quilliam, 1966; Munger, 1971). The smaller Pacinian corpuscle are, according to Schmidt (1971), velocity detectors.

In addition to the literature concerned with a detailed description of the receptors, there are also papers attempting to summarise existing knowledge (Andres, 1969; Munger, 1971; Poláček, 1971). Poláček (1971) goes farthest with this synthesis. He maintains that the nerve endings are working in groups. Every nerve terminal has several nerve endings, but always of the same kind. Poláček (1971) differentiates three basic forms of nerve endings in the entire body: "free" nerve endings, nerve endings with auxiliary cells and terminal corpuscles with inner cores. Differences in the sensory innervation of the different tissues consist of the varying occurrence of these three basic forms. Their occurrence depends

9

on the one hand on the phylogenetic development, and on the other hand on the function of the tissues concerned. In other words, the phylogenetic position determines the basic form of the terminal whilst the number of terminals depends on the function. An his example, Poláček (1971) used the innervation of the articular capsule, which may show different numbers of terminals in animals of phylogenetically similar positions. Nerve endings are fewer in the articular capsules of rarely flying birds (domestic fowl, domestic goose) than in birds spending more time on the wing (swallow, seagull, pigeon etc.) where we most frequently find corpuscles with an inner core. The earlier classifications were determined by the staining processes and resolution of light microscopy. The lack of sharpness in the outlines suggests a large number of shapes. With the introduction of the electron microscope, structural criteria became better defined. We did not consider it advisable to follow the method of other authors, of classifying end-organs studied by electron microscopy by the criteria of light microscopy.

We have attempted a new approach to the problem of end-organs and concentrated on the following matters:

1. Structure of the nerve endings and their classification according to structural characteristics in different mammals.

2. Dependence of the innervation on the structure of the epidermis and dermis (structure of the basal-cell layer and surface contour of the epidermis).

3. Organisation pattern of the receptors in the skin and their arrangement in receptor groups.

In order to solve the first problem, it was necessary to re-examine our knowledge to date, and we have attempted, on the basis of examinations and checks, to simplify the complicated classification of the mechanoreceptors. Our classification is based primarily on the ultrastructure of the nerve endings. We consider the differences in size and shape to represent adaptations to different types of skin surface and dermal structure.

The second problem, which is closely bound up with the classification of the nerve endings, is, in our view, the more important. The sensory nerve endings are directly dependent on the form and structure of the skin. The skin receives mechanical stimuli (e.g. pressure) and transmits them to the nerve terminals. The form of the basic types of the nerve terminals is adapted to the environment.

The third problem concerns the arrangement of individual types of receptors in receptor complexes. Each complex defined against the environment and generally contains the three basic types.

For the sake of completeness, our work contains a chapter on the innervation of hairs and sinus hairs (vibrissae). In these skin-related tissues, the arrangement of the receptor complexes is visible at a glance.

Material and Methods

The animals examined were: mole (*Talpa europaea*)—glabrous nasal skin, paw skin and sinus hairs; cat (*Felis silvestris f. catus* L.)—nasal skin, paw skin and sinus hairs; dwarf pig (*Göttinger miniature pig*)—snout skin and sinus hairs; monkey (*Macaca mulata, Macaca cynomolgus*)—glabrous digital skin of the volar side of the distal phalanges.

All the animals were anaesthetised with Nembutal (80 mg/kg intraperitoneal). The thorax was opened to expose the heart and pericardium. The left ventricle was incised and a tube inserted into the ascending aorta. The animals were perfused with a glutaraldehyde solution (6% buffered to pH 7.2 with 0.05 M phosphate). The perfusion lasted for 5–30 min, depending on the size of the animal. The mole and the cat required 5–15 min, the miniature pig and the monkey 15–30 min. The skin was cut into cubes with an edge length of 1–2 mm. The total fixation in the glutaraldehyde solution lasted one hour. This was followed by 2-hour post-fixation in a 1.0% OsO_4 solution in 0.1 M phosphate buffer solution with a 1% addition of sucrose, at 2–4°C. Dehydration was carried out in an alcohol series by the method of Luft (1961), and the material was embedded in Epon 812 (Serva).

An ultramicrotome (OmU2—Reichert and Porter Blum) was used to prepare semi-thin sections for light microscopy and ultra-thin sections for electron microscopy. The cutting planes were parallel and perpendicular to the skin surface. The semi-thin sections were stained by the method of Ito and Winchester (1963)—1 part pyronin to 3–4 parts toluidine blue. The ultra-thin sections were treated with uranyl acetate (1% in 1% acetic acid) for a total of 30 min, and lead citrate, by the method of Reynolds (1963). Electron microscope: Philips 300, 60 kV.

For the scanning electron microscope, we used a modified method based on Horstmann (1957). The skin samples were immersed in a 1% solution of acetic acid. After 12–24 hrs, the epidermis was separated from the dermis. Both tissues were stretched on a cork plate with pins and fixed in a 6% glutaraldehyde solution in 0.05 M phosphate buffer solution for 10 to 24 hrs. After fixation, the preparations were freeze-dried and fitted with the aid of the silver into the preparation holder for scanning electron microscopy. The preparations were shadowed with carbon and gold. Scanning electron microscope: Cambridge an Jeol.[1]

Results

A. Epidermal Nerve Endings

The nerve endings in the epidermis take two forms: 1. Free nerve endings. 2. Nerve endings with specialised cells (Merkel endings).

1. Free Nerve Endings

The free nerve endings were observed chiefly in the glabrous nasal skin of the mole. These endings were rarely found in the nasal epidermis of the cat, and not at all in the pig snout or the digital skin of the rhesus monkey.

The epidermal cones in the nose of the mole contain two types of endings (Fig. 1a–d): a) 1–2 thicker axons run in the centre of the cone; their diameter

Fig. 1a Glabrous nasal skin of the mole. Semi-thin section parallel to the skin surface (×800). A nerve fibre runs through the centre of each epidermal cone, and, at the distance of one prickle-cell, 20–22 nerve fibres (↑)

Fig. 1b Glabrous nasal skin of the mole. Section parallel to the skin surface (×11000). The nerve terminal (∗) contains accumulations of mitochondria and vesicles

Fig. 1c Glabrous nasal skin of the mole. Section perpendicular to the skin surface (×12500). The nerve fibre shows terminal thickening (1) and contains a large number of mitochondria and vesicles. It is situated between prickle-cells of epidermis (2)

Fig. 1d Semi-diagrammatic representation of the free nerve endings in the epidermis of the cone skin in the mole. Cross section: Free nerve endings with terminal expansions in the granular layer of the epidermis (1). The afferent non-myelinated nerve fibres (2) are enveloped by a Schwann cell. The basal lamina of the Schwann cell fuses with that of the epidermis (∗). Protrusion of the blood sinus (3). Horizontal section: The axons are invaginated in keratinocytes of the epidermal cones (↑) but continue in extracellular form

1 I wish to thank Cambridge of Dortmund and Jeol (representatives Contron in Hamburg) for making the scanning electron microscopes available to me.

For the drawings I am gratefully indebted to H. Hess of the Anatomical Institute Hamburg.

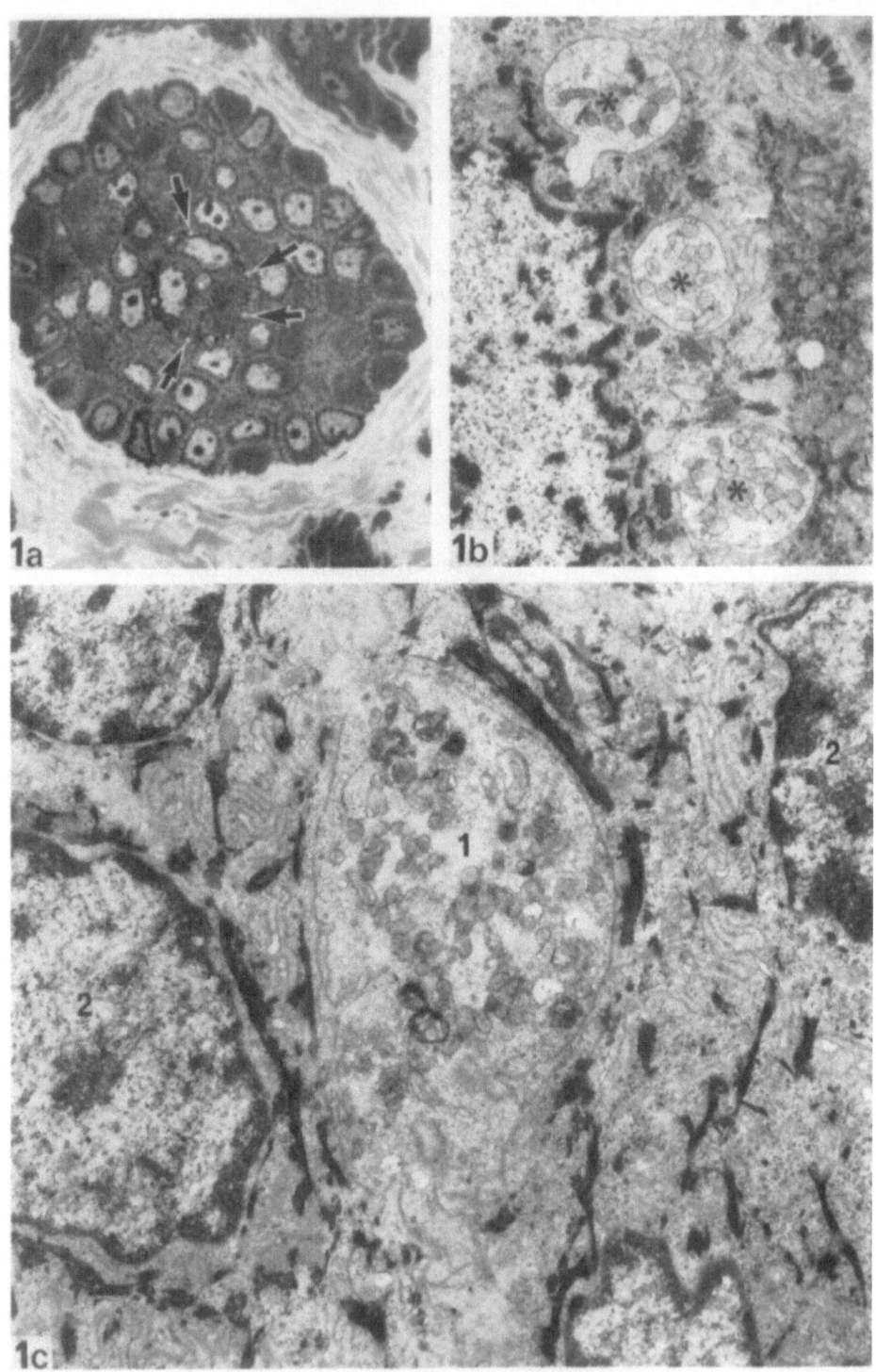

(Legend see p. 11)

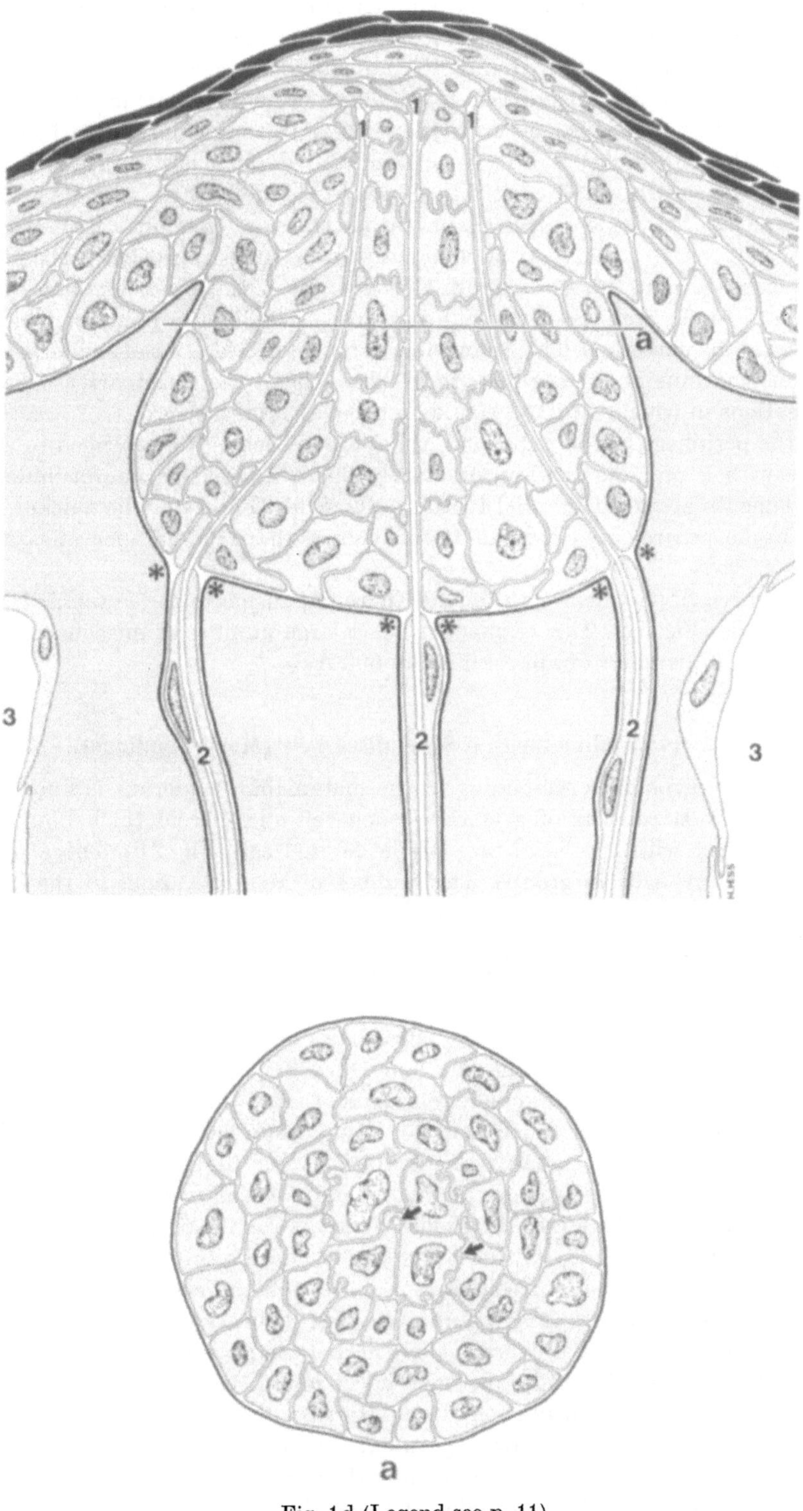

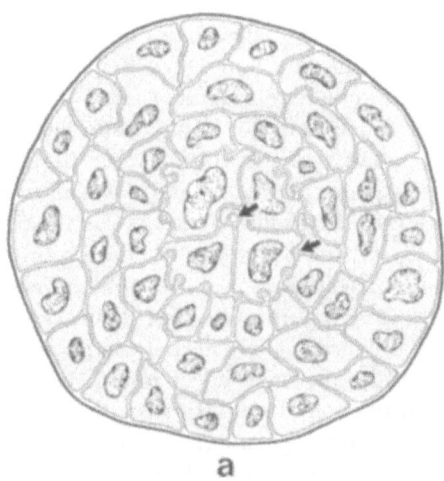

Fig. 1d (Legend see p. 11)

fluctuates between 2 and 2.4 μ. b) Thinner fibres with a diameter of 1–1.2 μ, running at a distance of 12–15 μ (stratum spinosum–cell diameter) from the centrally positioned fibres. They number about 20. Their course is parallel to that of the centre fibre and in cross sections shows them to be arranged in a circle (Fig. 1 a). Both types of axon run perpendicular to the epidermal surface. They represent the continuation of axons from the dermal nerve plexus situated underneath each epidermal cone. The fibres in the dermal nerve plexus are not myelinated and have no perineurial sheaths. They are enveloped by a Schwann cell and its basal lamina After short distance, they approach perpendicular to the basal lamina of the epidermale cone. Both the basal laminae—of the epidermal cone as well as of the Schwann cell—fuse at the point at which the axons enter the epidermis. The Schwann cell too ends here. In the epidermal cone, the axons continue in extracellular form. The cells of the germinative zone form invaginations in which the axon run as in grooves (Fig. 1 c).

At the periphery of the axoplasm, there are mitochondria arranged in a ring, vesicles with a diameter of approximately 600 Å as well as neurotubules and neurofilaments arranged parallel to the nerve fibre (Fig. 1 b). The thicker nerve fibres in the centre are richer in mitochondria than the thinner ones at the periphery.

Each nerve fibre, thick or thin, ends in an expansion in the granular layer of the epidermis (Fig. 1 c). This expansion has a large number of mitochondria and vesicles with a diameter of approximately 600 Å.

2. Nerve Endings with a Specialised Cell (Merkel endings)

The second type of nerve ending in the mammalian epidermis is known as a Merkel ending. It consists of a Merkel touch cell and discoid thickening of the axon in contact with the basal area of the Merkel cell (Fig. 2 b). These endings are generally arranged in groups. The number of Merkel endings in the base of the epidermal cones are: 3–5 in the mole (Fig. 2 h) 5–15 in the cat, and in the miniature pig 10–20 (Fig. 2 a). The digital epidermis of the monkey contains groups of 5–10 such endings in the glandular ridges where these are pierced by the gland ducts.

Merkel endings were found in the sinus hairs of all the animals examined. The endings formed a cuff round the hair (Figs. 9 c, d) which is situated below the sebaceous gland. The Merkel endings number more than 600.

The specialised Merkel touch cell (Fig. 2 b) differs clearly in structure from the other cells of the germinative zone. It is larger (9–16 μ in diameter) and ellipsoidal. The nucleus is large and lobed. The cytoplasm is light and has a large number of filaments which run in various directions and continue into the digitate processes of the cytoplasm of the Merkel cell. The digitate processes differ in length and enter into the intercellular spaces as well as into indentations of the cytoplasm of the cells of the germinative zone (Fig. 2 d). Their position and structure is reminiscent of anchorage. The Merkel cell is connected with keratinocytes by desmosomes (Fig. 2 d). We have never found such desmosomes between digitate processes and keratinocytes. In addition to filaments, the Merkel cell has a well developed Golgi apparatus, granular endoplasmic reticulum, ribosomes which are partially free but partially arranged in rosettes, and glycogen granules. The occurrence of osmiophilic granules (diameter 800–1 000 Å) is a typical feature of the Merkel cell. They are situated in that part of the cytoplasm which is in contact with the thickening of the nerve. Very occasionally pigment granules are found in the cytoplasm. By contrast with the above granules, these have a diameter of 0.1 to 1.0 μ. They are probably melanin granules. It is relatively easy to distinguish between melanocytes and Langerhans cells. Neither of these (Fig. 2f, g) has desmosomal links with the epidermal cells, and no contact with nerve fibres.

The Merkel cells of the sinus hairs have the same structure as those in the glabrous skin, but their shape is different. They form longish oval discs which are arranged in scales in the hair follicle of the sinus hair.

The afferent nerve fibre forms a discoid or cup swelling at the Merkel cell (Fig. 2b, c). It is myelinated in the dermal plexus under the epidermal cone, where it generally has an endoneurial or perineurial sheath. The endoneurium consists of largely longitudinal collagen fibres interspersed with flat fibrocytes. The perineurium is formed by 1–2 layers of flat cells.

At the last node of Ranvier, the axon may form branching terminals. The Schwann cell loses its myelin sheath 10–50 μ under the epidermis. Enveloped by the Schwann cell and its basal lamina, it then approaches the basal lamina of the epidermal cone. At the site where the axon penetrates the epidermis, the

Fig. 2a Snout skin of miniature pig. Section perpendicular to skin surface (×1200). The epidermal cone contains a group of approx. 15 Merkel endings (↑). Encapsulated corpuscles with inner core (∗) are situated below the cone

Fig. 2b Hairless nasal skin of the cat (×11000). The base of the Merkel cell (1) lies close to a nerve disc (2). The cytoplasm of the Merkel cell forms digitate processes (↑)

Fig. 2c Glabrous nasal skin of the mole. Section parallel to skin surface (×8500). The cytoplasm of the Merkel cell (1) is surrounded by a nerve disc (2)

Fig. 2d Epidermal cone in the glabrous nasal skin of the mole (×23000). The cytoplasm of the Merkel cell forms digitate processes (∗) extending into the intercellular spaces of the prickle-cell layer. Between the processes, the membrane of the Merkel cell is linked by desmosomes with the cells of the stratum spinosum (↑)

Fig. 2e Glabrous nasal skin of the cat. Detail of a synapse-like structure between nerve disc and Merkel cell (×40000). At the site of contact, the membrane of the axolemma (∗) is thicker than that of the Merkel cell. Presynaptically, the Merkel cell (1) contains osmiophilic granules and the nerve disc (2) contains vesicles

Fig. 2f Glabrous nasal skin of the cat. Section perpendicular to the skin surface (×9000). Difference between a melanocyte and Merkel cell. Merkel cell with desmosomes (↑). A melanocyte (2) lies close to the Merkel cell (1). Unlike the Merkel cell, the melanocyte has no desmosomal links with the cells of the stratum spinosum

Fig. 2g Glabrous nasal skin of the cat (×11000). The Langerhans cell in the prickle-cell layer of the epidermis has no desmosomal links with the prickle cells; it contains no pigment granule

Fig. 2h. Semi-diagrammatic representation of Merkel nerve ending in the epidermis of the nasal cone skin of the mole. Cross section: Merkel cells (1) are situated in the base of the cone. The nerve discs (2) are contiguous with the basal areas of the Merkel cells. In the deeper layers of the dermis, the nerve fibres are myelinated (4) but immediately below the cone they are enveloped only by Schwann cells (3). As the axon enters the epidermis, the basal lamina of the Schwann cell fuses with that of the epidermal cone (∗). Extroversion of the blood sinus (5). Horizontal section: The Merkel cells (1) extend digitate processes between the keratinocytes. At (2), the nerve disc is incised. (↑) incised axon of the free nerve endings

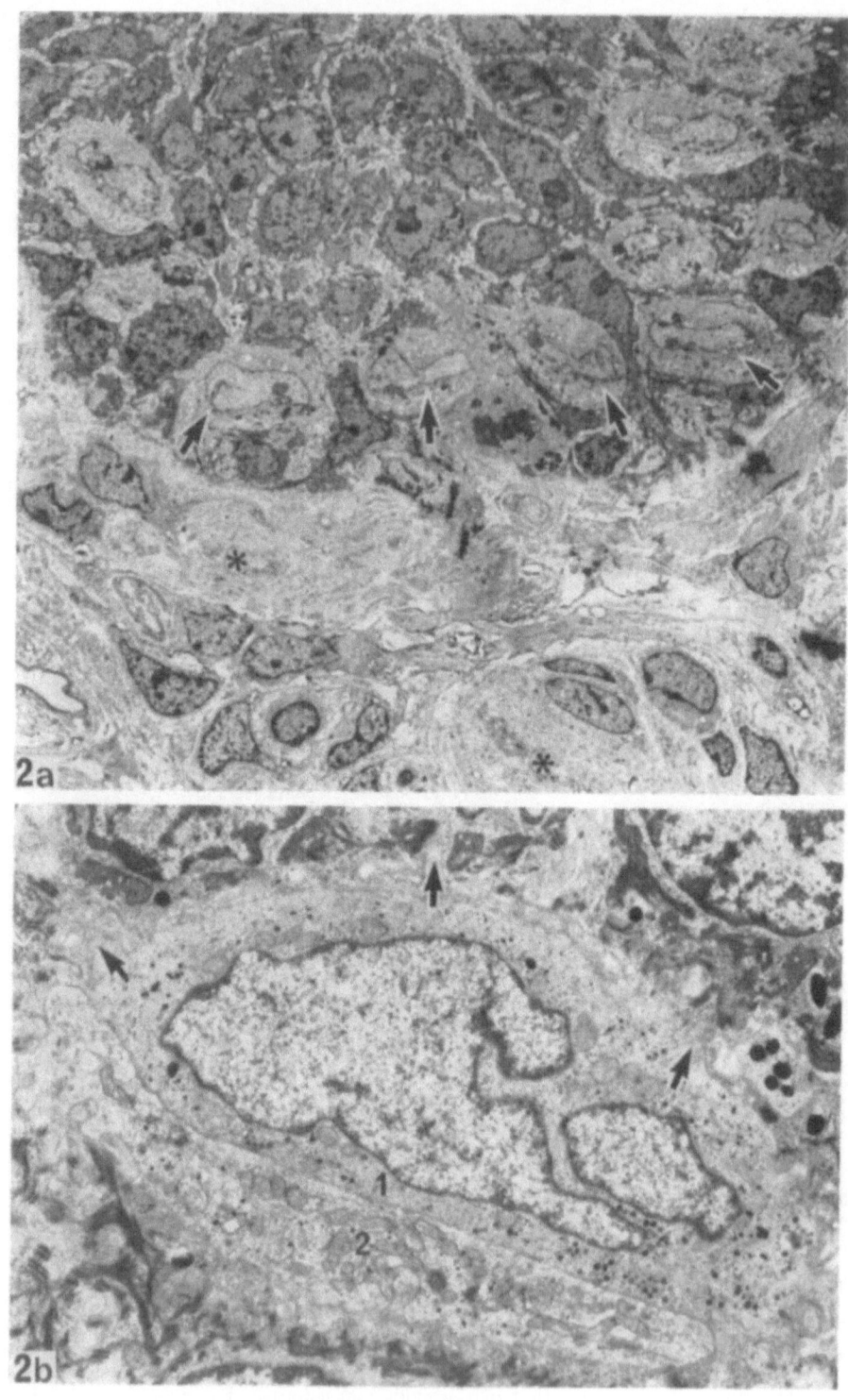

(Legend see p. 15)

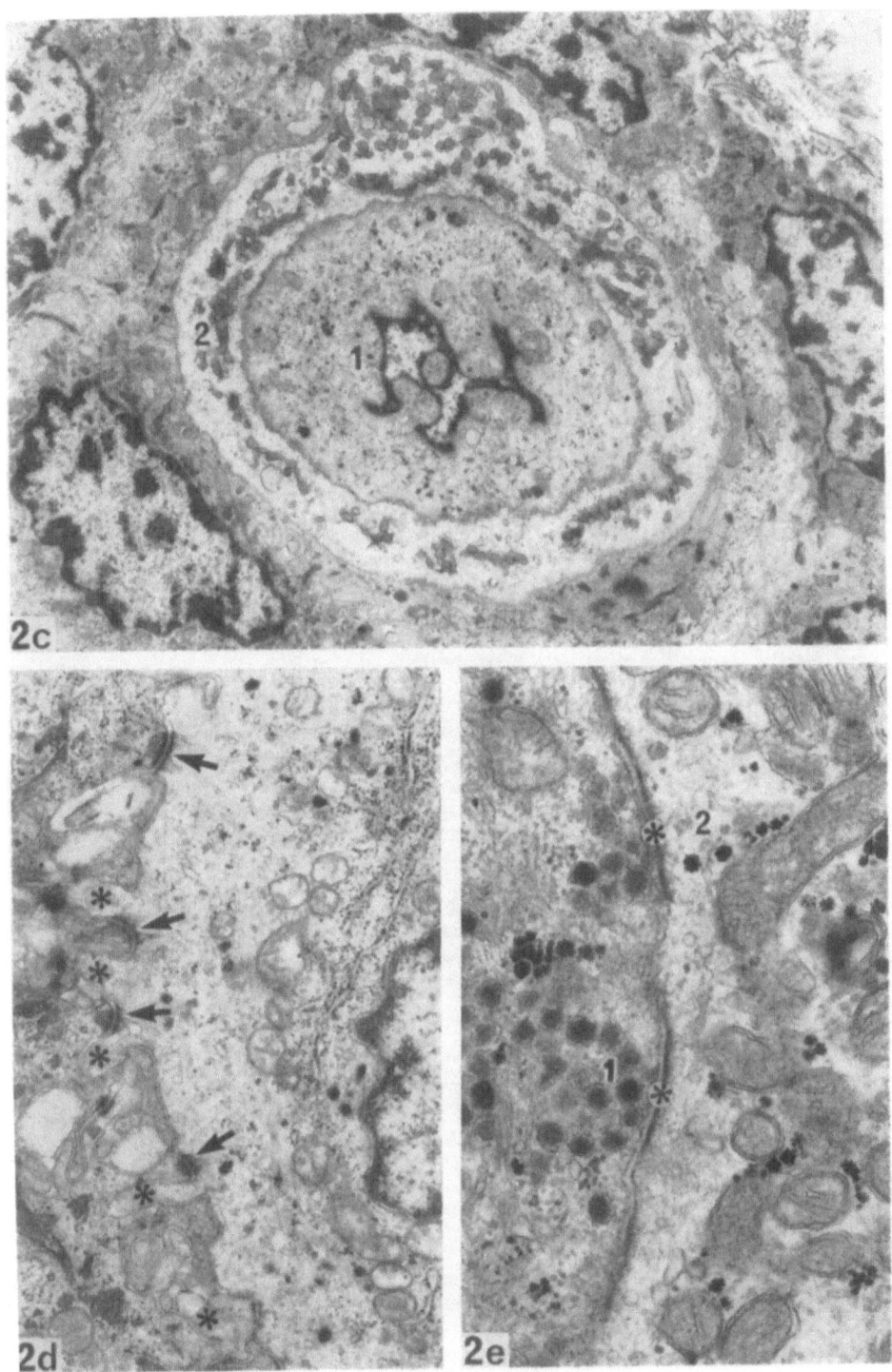

2c

2d

2e

(Legend see p. 15)

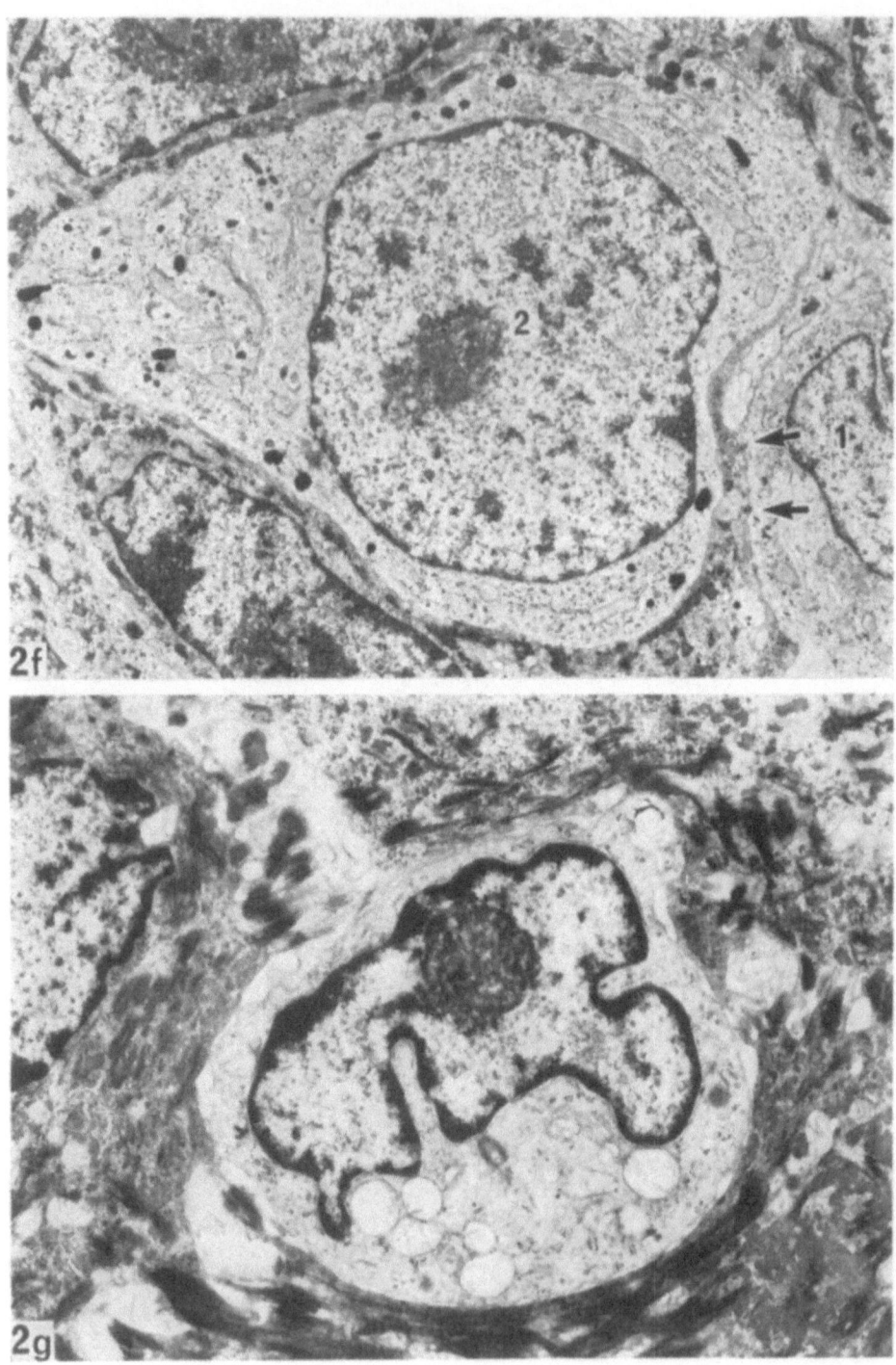

(Legend see p. 15)

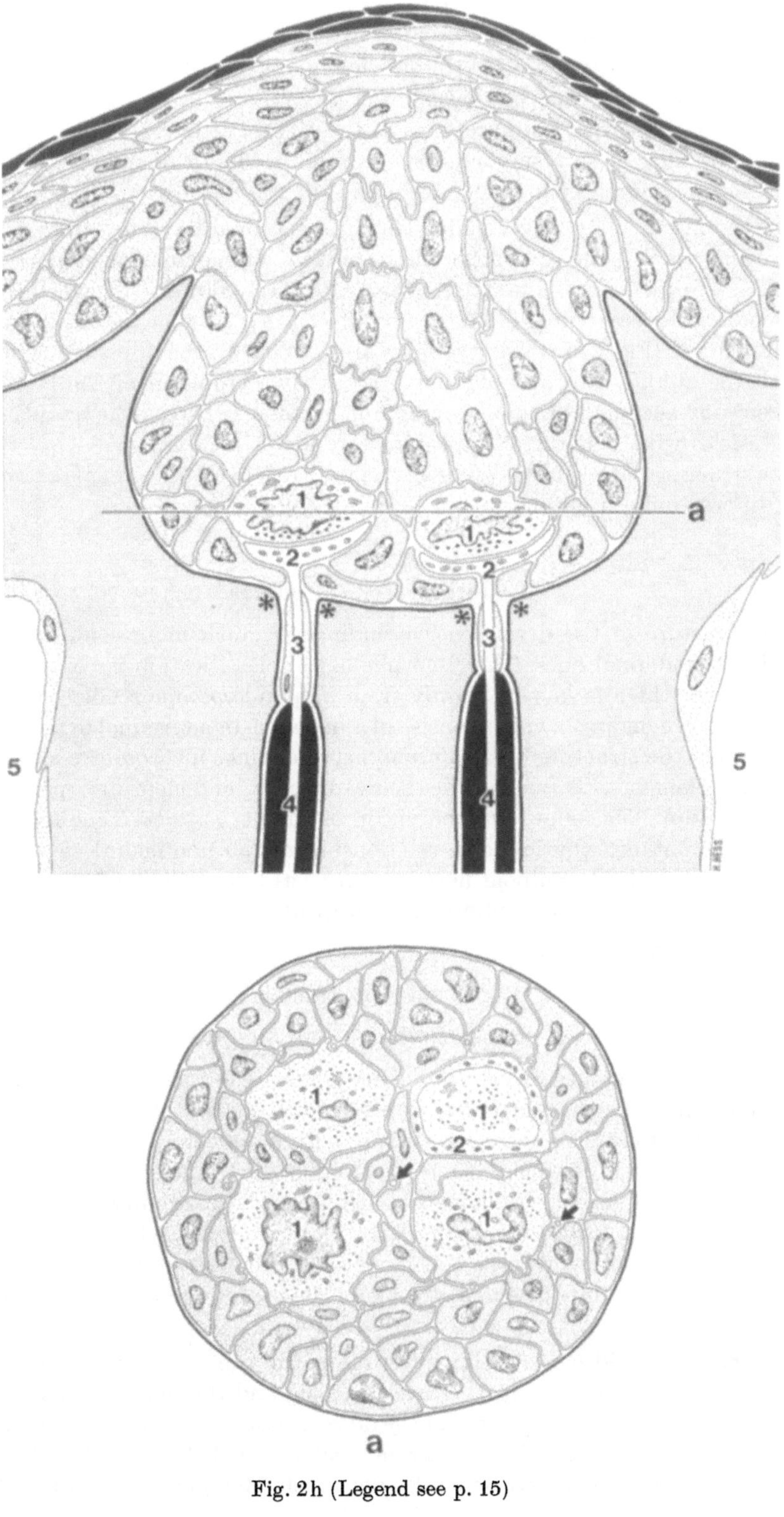

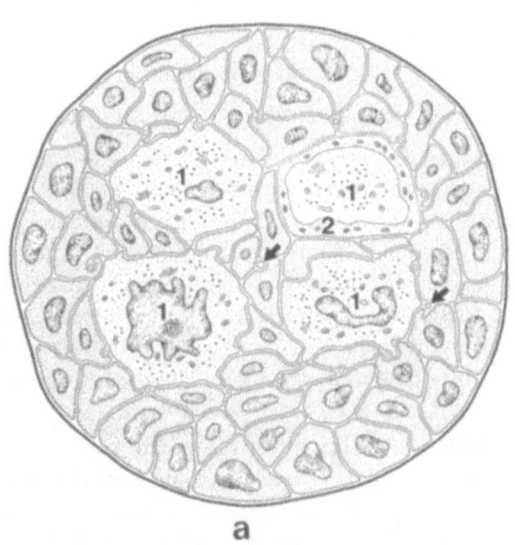

Fig. 2h (Legend see p. 15)

basal lamina of the Schwann cell fuses with that of the epidermal cone in a manner similar to that of the free nerve endings (Fig. 9e). This may be the site of the next series of branching ramifications.

After a short progress, the axon forms the discoid or cup thickening which is in direct contact with the basal area of the Merkel cell (Fig. 2b, c). Cells of the germinative layer closely surround the axon as well as the area of the swelling that is not in contact with the Merkel cell.

The axoplasm of the nerve fibre and the terminal swelling contains neurotubules and neurofilaments. The swelling contains a large number of mitochondria as well as vesicles with a diameter of approximately 600 Å. In several cases we saw synapse-like structures in the region of contact between the nerve disc and the Merkel cell (Fig. 2e). Here swellings are formed by the opposing membranes of both the Merkel cells and the nerve disc. The cytoplasm of the Merkel cells here contains accumulations of osmiophilic granules whereas the axoplasm contains vesicles with a diameter approximately 600 Å (Fig. 2e).

The structure of the Merkel endings was the same in all the animals examined. They differed only in number.

B. Dermal Nerve Endings

The structure of the dermal nerve endings is much more complicated than that of the epidermal ones. Their morphological classification is complicated by their variety, which is known mainly from light microscope studies. Unlike the epidermal nerve endings, which consist of a free axon or a terminal expansion with a Merkel cell, the structure of the dermal nerve endings incorporates a number of non-neural elements. These include Schwann cells, endoneurium, perineurium, and epineurium. The following subdivision of the dermal nerve endings is based on the different distribution and relationship of the non-neural elements with the nerve terminals: 1. Bulboid nerve endings: a) Simple bulboid nerve endings. b) Dendritic bulboid nerve endings; 2. Encapsulated corpuscles with inner core.

1. Bulboid Nerve Endings

a) Simple Bulboid Nerve Endings

In order to differentiate between a nerve terminal and continuous axon, we must first define the characteristics of a terminal. Diagnostic features consist of accumulations of mitochondria and vesicles with a diameter of 600 to 800 Å, and the terminal is generally thicker than the afferent fibre. The body of the Schwann cell is drawn out to form a thin lamella which invests the terminal. The membrane of the Schwann cell is studded with a large number of pinocytotic vesicles in this area. All the axons having these features are classified as bulboid nerve endings, and all others as continuous axons ending as Merkel endings or free nerve endings in the epidermis.

No simple bulboid nerve endings were found in the nasal dermis of the mole. In the hairless skin of the cat, the miniature pig and the monkey, these endings were found in the stratum papillare of the dermis close to the epidermis (Fig. 3a, b). The nerve fibres originate in the dermal plexus which lies about 200 to 400 μ below the epidermis. They are generally not myelinated and, covered by a lamella from the Schwann cell, spiral to the skin surface. These axons do not possess endoneurial, perineurial, or epineurial sheaths. The axon expands approximately 2–10 μ below the basal lamina of the epidermis (Fig. 3a). Such expansions are invested with a lamella of the Schwann cell and its basal lamina. In some places the thickened axons are no longer covered by the Schwann cell and are enveloped solely by the basal lamina. The longitudinal axis of the nerve endings is predominantly parallel to the surface of the basal-cell layer of the epidermis.

b) Dendritic Bulboid Nerve Endings

Unlike the simple bulboid nerve endings of the dermis, the dendritic bulboid endings form palisades, spirals and coils. We found these endings in the hairs and vibrissae of the mole, cat and dwarf pig and in the glabrous skin of the monkey. The type associated with hairs is generally called a palisade nerve ending, that of vibrissae, a lanciform nerve ending, and that in the ridged skin of the monkey, Meissner's corpuscle.

Dendritic Nerve Endings of Hairs and Vibrissae. These endings occur in the connective tissue sheath about the hair follicle or in the hair follicle of sinus hairs. The afferent fibre is myelinated. The endings run parallel to the hair without endoneurial, perineurial, or epineurial sheaths. In the last node of Ranvier, the axon may ramify repeatedly. After losing its myelin sheath, the axon thickens. In cross section these endings form a flattened oval. They lie between two lamellae of a Schwann cell (Fig. 4a, 9g–j, 10a). These lamellae contain a large number of pinocytotic vesicles, on the inner surface turned towards the nerve terminal, as well as on the outer surface. The terminal appears oval in cross section; the narrow sides may have digitate processes. At this point the Schwann cells have no lamellae so that the filaments of nerve terminal are covered only by the basal lamina of the Schwann cell. Sometimes the basal lamina fuses with that of the hair or vibrissae. The terminal is then in direct contact with the basal lamina of the basal-cell layer (Fig. 9g, h). Occasionally two or three axon terminals are enclosed in one Schwann cell.

The total number of lanciform endings associated with one vibrissa is around 200, and for normal hair about 100. The number depends on the hair diameter.

The differences in the size and arrangement between the lanciform endings of normal hair and sinus hair (vibrissae) will be discussed more fully in the chapter on "Mechanoreceptor Complexes".

Dendritic Bulboid Nerve Endings in the Ridged Skin of the Monkey. These endings are generally called Meissner's corpuscles. They occur in the stratum papillare of the dermis, occupying crypts in the epidermis within the papillae (Fig. 4b). Their longitudinal axis is perpendicular to the epidermal surface; their length is approximately 100 μ and their width about 50 μ (Fig. 4d).

The afferent axon, which has a diameter of approx. 4–8 μ, is myelinated in the nerve plexus under the dermal papillae and invested with endoneurium and perineurium. The perineural sheath consists of 1–3 layers having the same structure as the perineurium in the nerve plexus. One to three such axons may enter one dermal papilla. Here the axon loses its endoneurial and perineurial sheaths and finally also its myelin sheath, after which it forms several terminals (Fig. 4b, c). These spiral towards the basal-cell layer of the epidermis. They are covered by

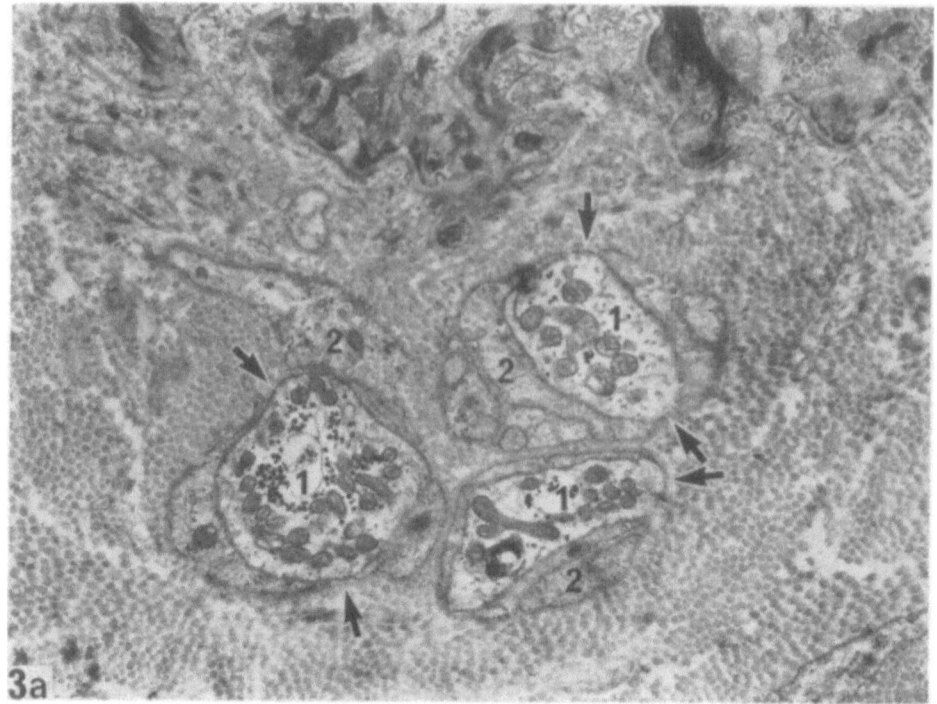

Fig. 3a Simple bulboid nerve endings in the nasal dermis of the cat. Section perpendicular to the skin surface (×11000). The nerve endings (1) have numerous mitochondria and vesicles and are invaginated in Schwann cells (2). In some places the cytoplasm of the Schwann cell is absent and the endings are invested only with a basal lamina (↑)

Fig. 3b Semi-diagram of the simple bulboid nerve endings in the nasal skin of the cat. The terminal swelling of the axon (1) is invested with a Schwann cell and its basal lamina (2). Incision of the capillary bed (3). The detail represents a cross section through a simple bulboid nerve ending. The expansion of the axon contains numerous mitochondria and vesicles and is invested with a Schwann cell lamella and its basal lamina

lamellae of the Schwann cells, the membrane of which is studded with pinocytotic vesicles. The terminals (Fig. 4c) contain accumulations of mitochondria, as well as vesicles with a diameter of approximately 800 Å. In addition to these organelles, which are typical for nerve fibre terminals, we found central neurofilaments and, at the periphery, neurotubules. The dendritic bulboid terminals are situated perpendicular to the longitudinal axis of the Meissner corpuscle and consequently parallel to the skin surface.

Between the spirals of the axons and their terminal expansions, we found collagen fibres, also in spiral arrangement. The collagen fibres are interspersed with fibrocytes. The connective tissue portions may be regarded as continuations of the endoneurium.

On the outside, the Meissner corpuscle is enclosed by bundles of collagen fibres which adapt their form to that of the corpuscle. In some places we find fibrocytes and their flat continuations. Here we are clearly not dealing with a perineurial capsule but with a curved layer of connective tissue.

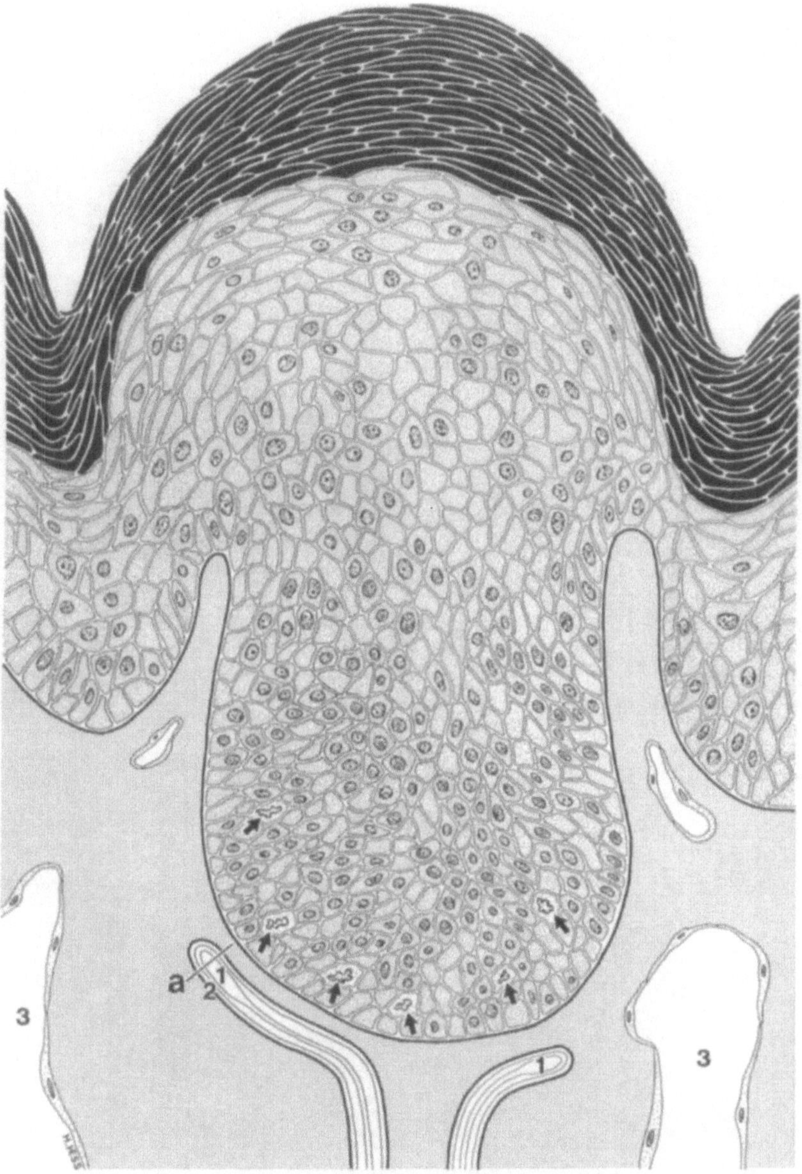

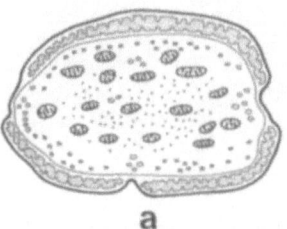

a

Fig. 3b (Legend see p. 22)

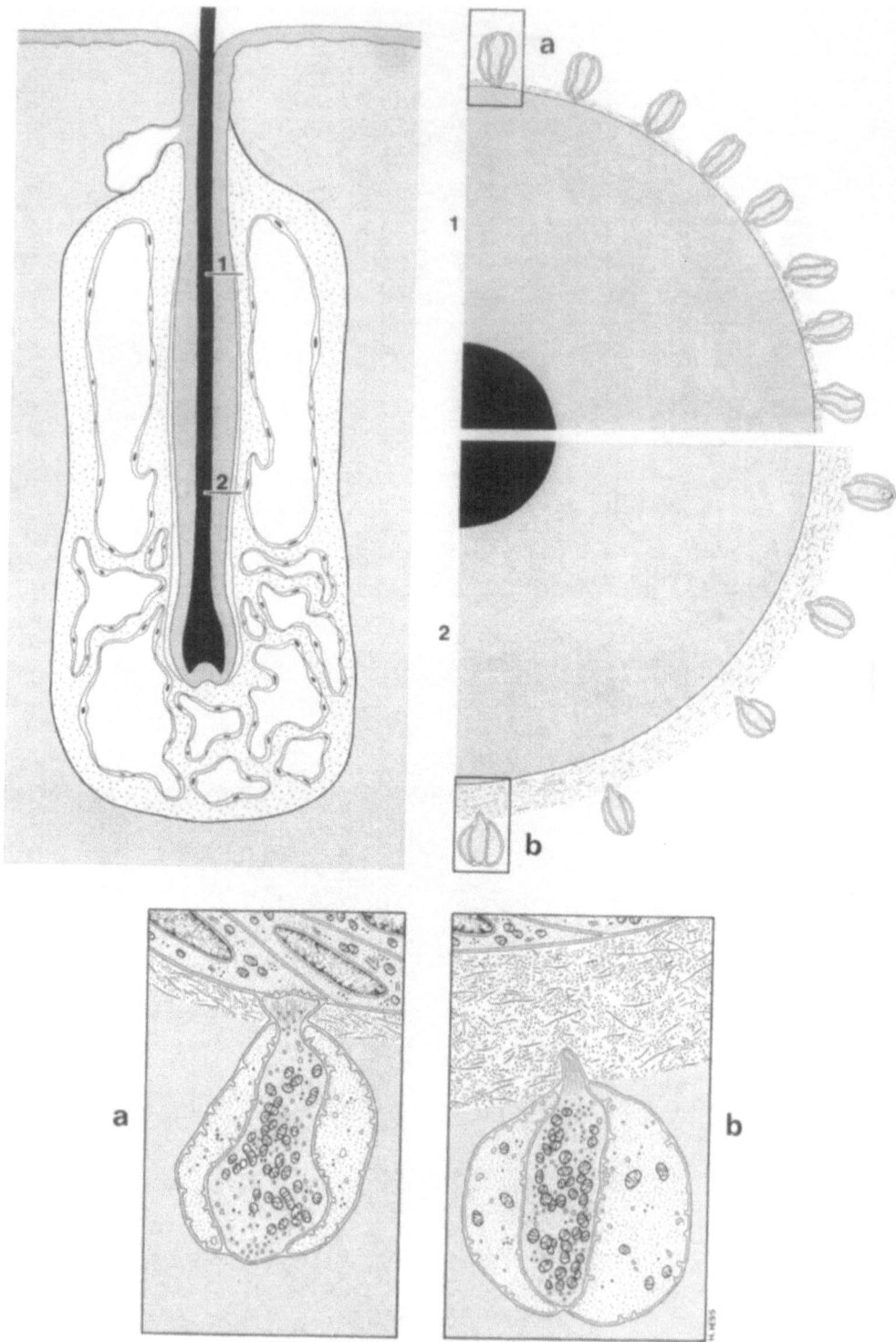

Fig. 4a (Legend see p. 27)

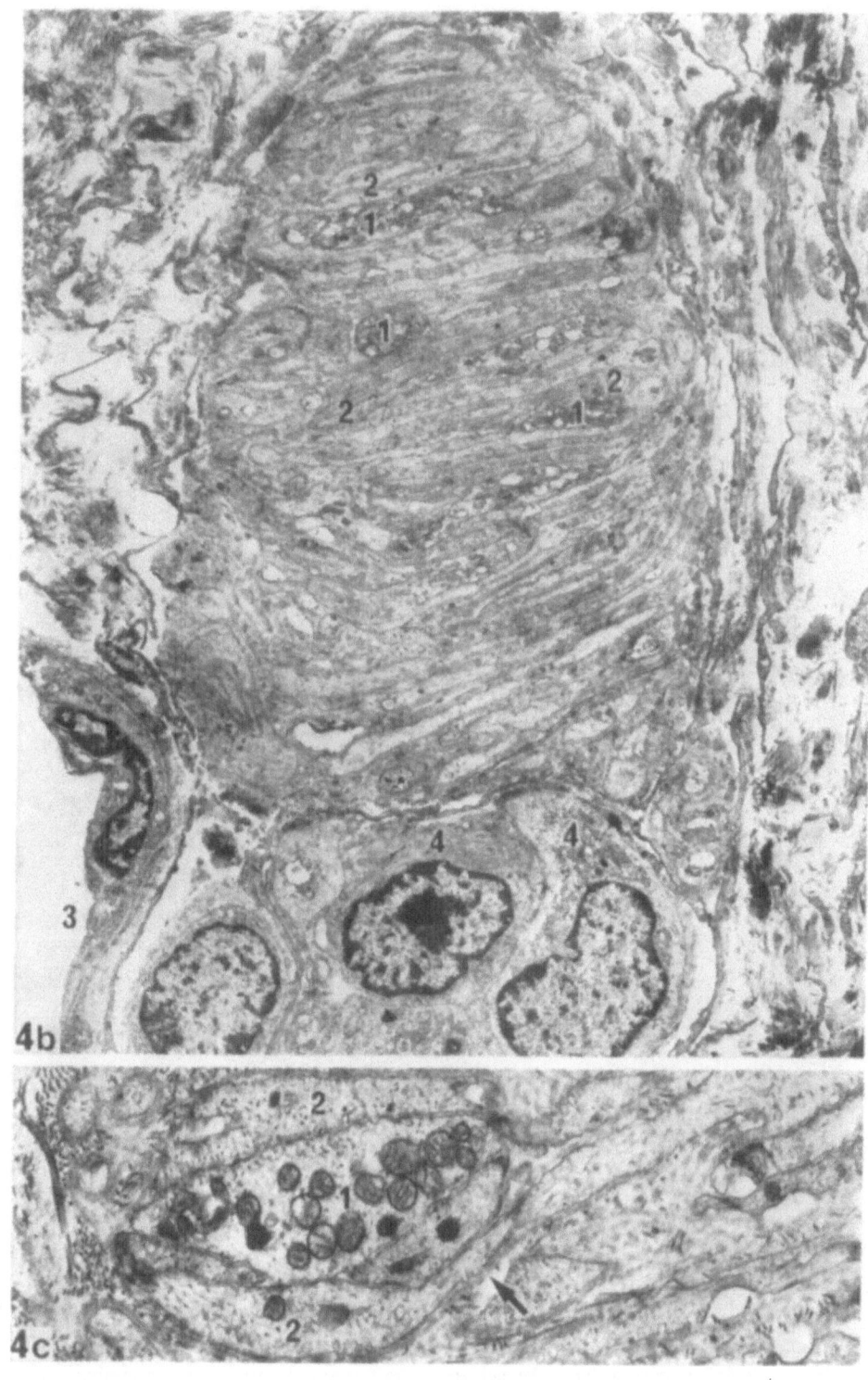

(Legend see p. 27)

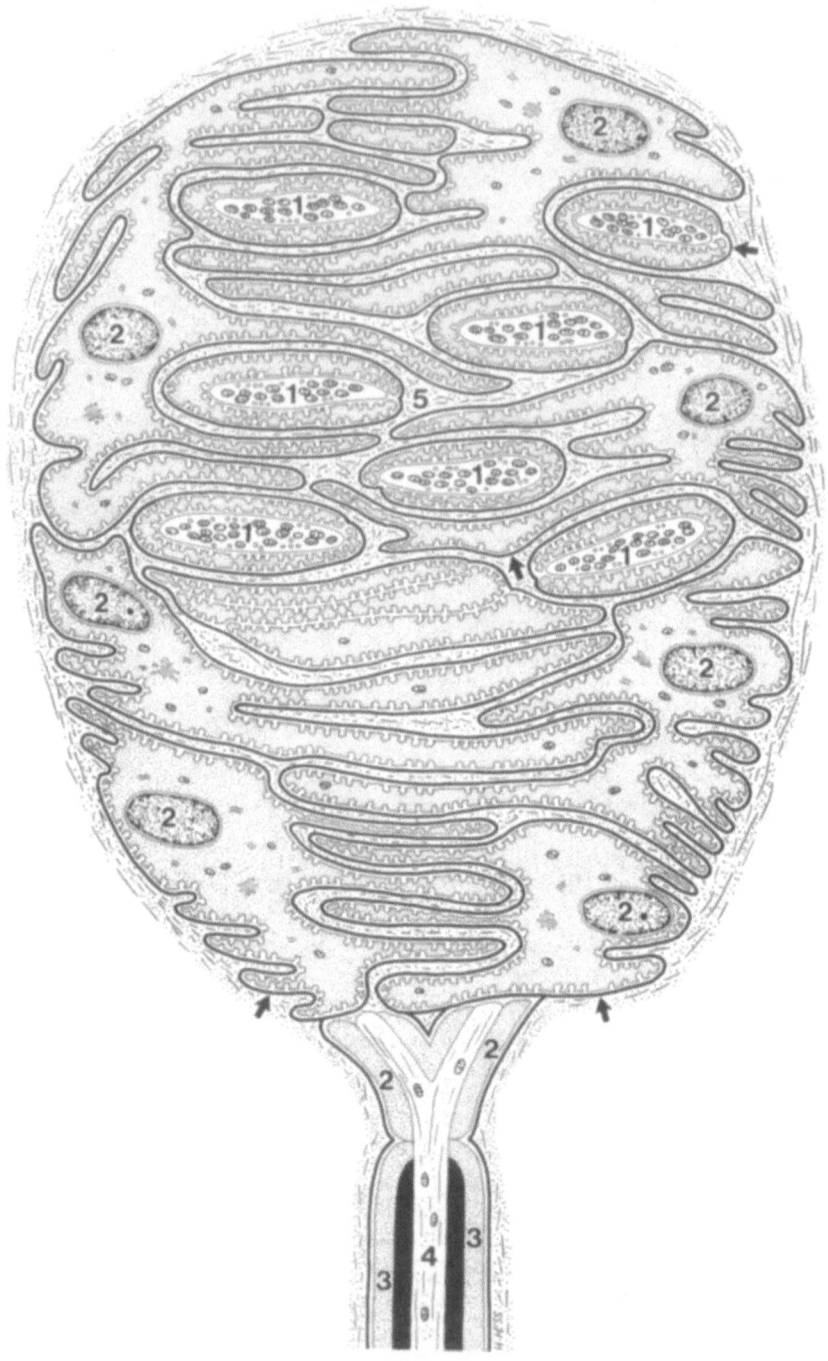

Fig. 4 d

Fig. 4d. Semi-diagrammatic representation of the dendritic bulboid nerve endings in the ridged skin of the monkey (Meissner endings). *1* Section of the spiral nerve terminals. *2* Schwann cells from a lamellar system. Their nuclei are on the external surface of the endings. The afferent axon (*4*) is myelinated (*3*). The lamellar system of the Schwann cells is enveloped by a basal lamina (↑). Collagen fibres occupy the spaces between the lamellae (*5*)

2. Encapsulated Corpuscles with Inner Core

This type of nerve ending is extremely variable. The corpuscles may vary in size and in the structure of their inner core. The thickness of the capsule too is variable. The end-organs consist of a nerve fibre, an inner core, a subcapsular space and a capsule.

The axon has three portions: the afferent myelinated segment, the middle segment which is inside the inner core, and the third bead-shaped part which ends in the distal part of the inner core (Fig. 5a–d, i).

The myelinated part of the axon has a diameter of approximately 3–8 μ. The thickness of the afferent axon depends on the size of the corpuscle. Smaller corpuscles have thinner axons, the large Pacinian corpuscles have thicker ones. These afferent axons are enveloped by a Schwann cell with a myelin sheath (Fig. 5d). The Schwann cell is separated from the endoneurium by a basal lamina. The endoneurium consists of spiral collagen fibres and fibrocytes. The perineural sheath consists of 1–5 layers of flat cells. Each layer is covered inside and outside with a basal lamina. The flat cells of the layers are linked by desmosomes. The basal laminae of the outer and inner layers are separated by gaps containing bundles of collagen fibres. The collagen fibres are cross-linked. On the outer side, the perineurium changes gradually into the connective tissue of the dermis. This connective tissue layer is normally described as epineurium.

The axons contain central neurofilaments and peripheral neurotubules. Mitochondria accumulate at the last node of Ranvier, which is inside the inner core (Fig. 5d). At this point the axon may fork, which is then followed by cleavage of the inner core.

The middle segment of the axon goes through the central part of the inner core (Fig. 5a, c). It has a diameter of approximately 2–5 μ. This part contains mitochondria in annular arrangements in the axoplasm as well as neurofilaments in the centre and neurotubules near the axolemma. In places the digitate processes of the axon may penetrate the spaces between the lamellae of the inner core; they contain vesicles with a diameter of approximately 600 Å. The length of this part depends on that of the inner core and consequently on the size of the corpuscle.

The terminal portion of the axon is expanded to form a bead which has a diameter of approximately 5–12 μ. These terminal expansions frequently extend processes, containing vesicles, between the lamellae of the inner core (Fig. 5b). The axoplasm of the extension contains accumulations of mitochondria and vesicles. In some parts of the terminal expansion there may be thickening of the

Fig. 4a Semi-diagram of the dendritic bulboid nerve ending of sinus hair. In the upper region of the hair follicle (1), the glassy membrane is thin. The digitate processes of the lanciform endings come in contact with the basal cell layer of the hair follicle (a). In the lower region of the hair follicle, the glassy membrane is thick, and the digitate processes do not reach beyond it (b)

Fig. 4b Meissner corpuscle from the digital skin of the monkey. Section perpendicular to skin surface (×4000). The bulboid nerve endings (1) run spiralling and are invested with lamellae from the Schwann cells (2). A group of Schwann cells (4) is situated at the foot of the corpuscle, surrounded by a perineurial cells. As a rule, the corpuscles lie close to capillaries (3)

Fig. 4c Detail of Figure 4b (×18000). The nerve fibre endings (1) are surrounded by lamellae from the Schwann cells with membranaceous vesiculations (2). The basal lamina (↑) invests the lamellae of the Schwann cell

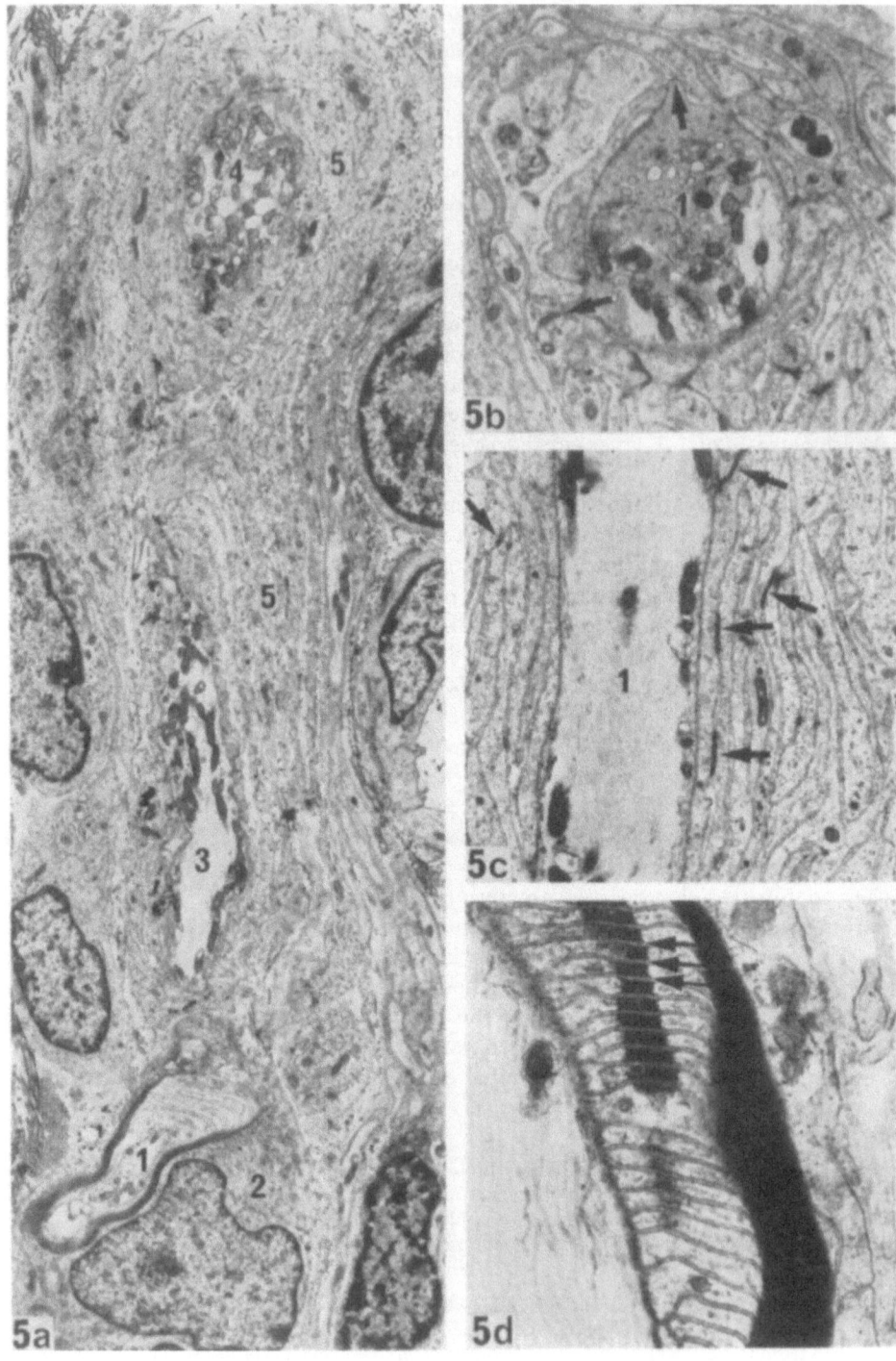

Fig. 5a. Glabrous nasal skin of the cat. Section parallel to the skin surface. Longitudinal section of simple corpuscle with inner core (×5000). *1* Myelinated nerve fibre; *2* Schwann cell; *3* nerve fibre in inner core; *4* bead-like thickening of nerve fibre; *5* lamellae of the inner core

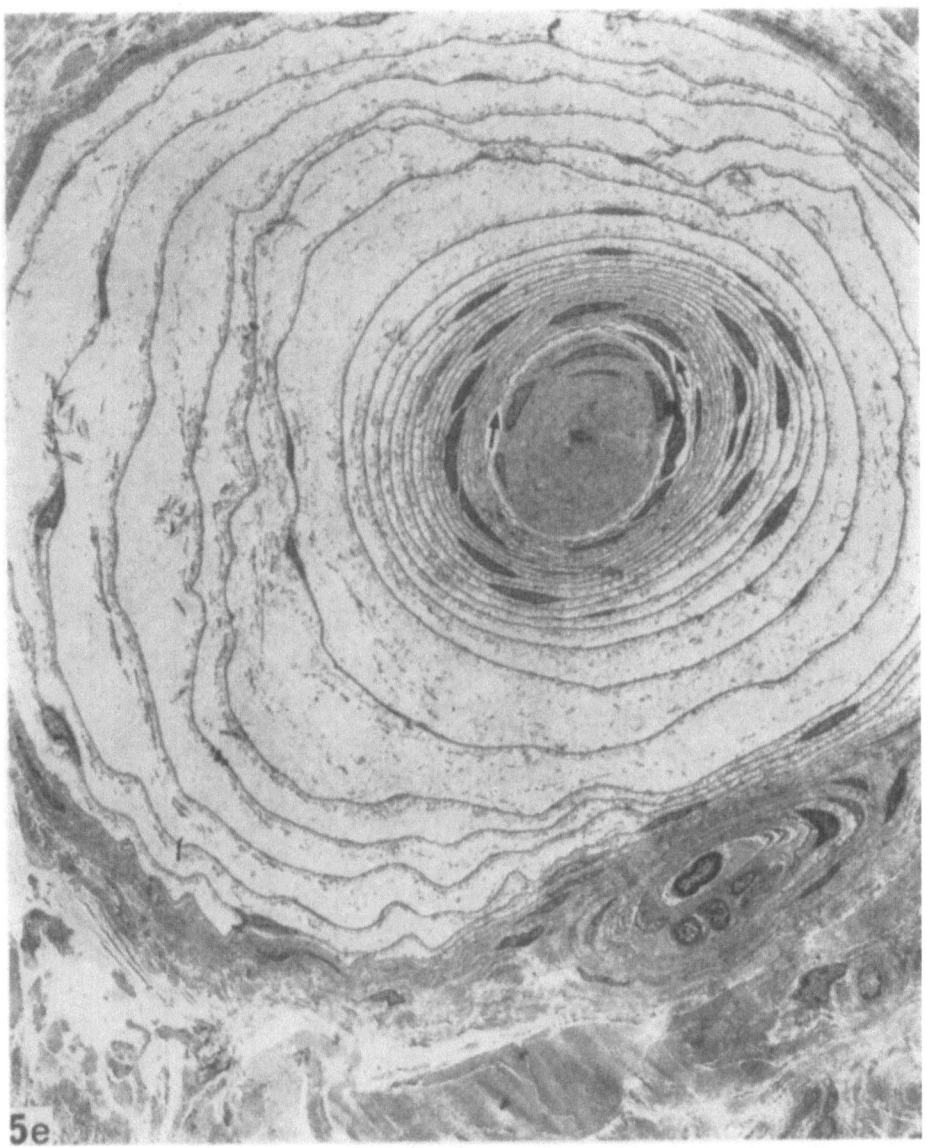

5e

Fig. 5b. Glabrous nasal skin of the mole. Section parallel to skin surface (×11000). Th distal part of the nerve fibre shows bead-like thickening (*1*). This swelling has digitate processes which extend between the lamellae of the inner core (↑)

Fig. 5c. Section through the central part of the inner core of a simple corpuscle from the glabrous nasal skin of the mole (×11000). A nerve fibre (*1*) runs inside the inner core and in addition to mitochondria and vesicles, contains central neurofilaments and peripheral neurotubuli. The lamellae of the inner core, which contain many pinocytotic vesicles, are linked in some places by desmosom-like structures (↑)

Fig. 5d. Ending of the myelin sheath in the inner core of the corpuscle (×27500). The cytoplasm lamellae of the Schwann cell are linked by desmosom-like structures (↑)

Fig. 5e. Glabrous digital skin of the rhesus monkey. Section perpendicular to the skin surface (×900). Vater-Pacini corpuscle with inner core. The inner core contains a nerve fibre. The subcapsular space is placed between inner core and the capsule (↑). The corpuscle is covered by a capsule (30 layers)

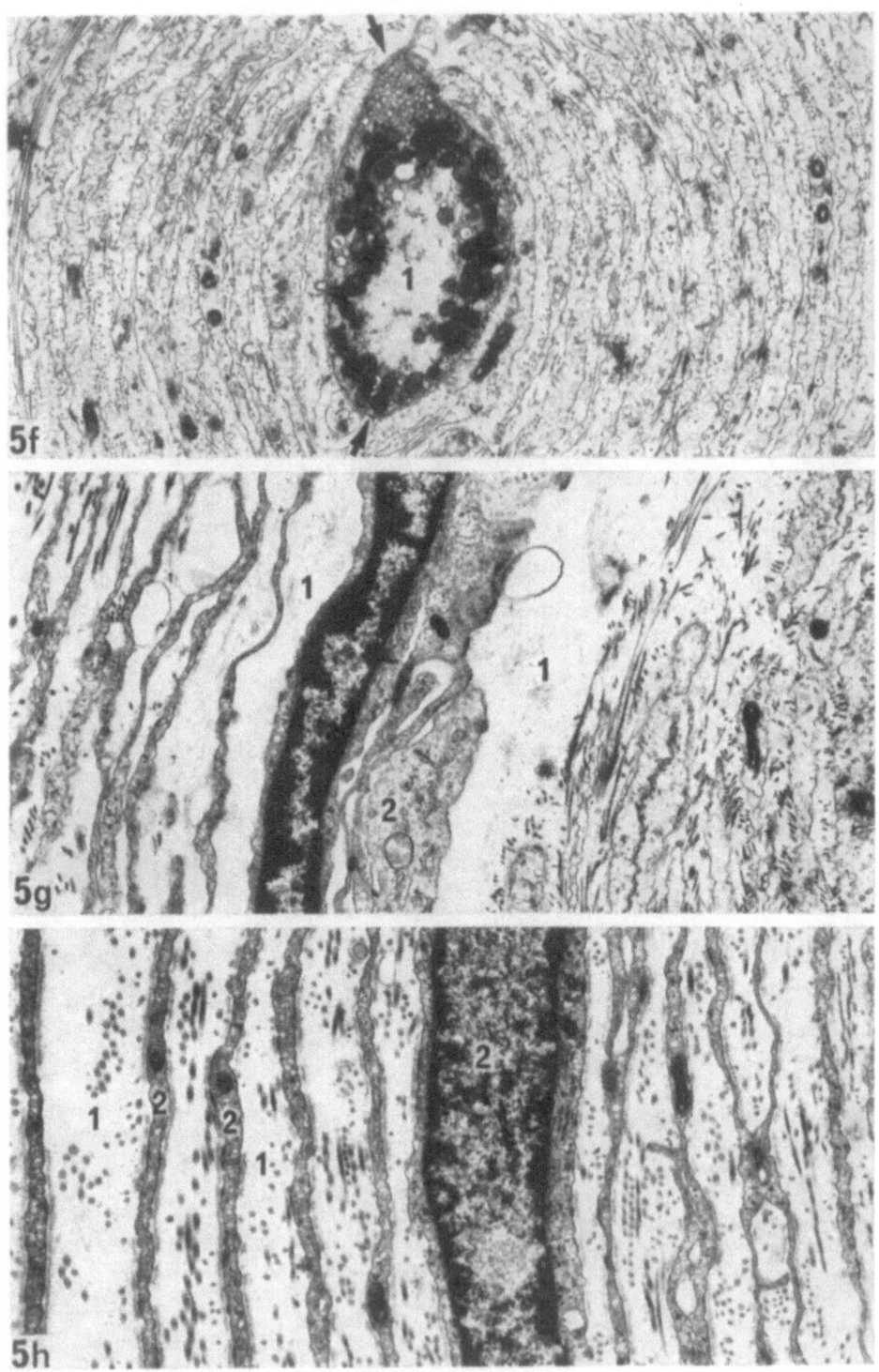

5f

5g

5h

(Legend see p. 32)

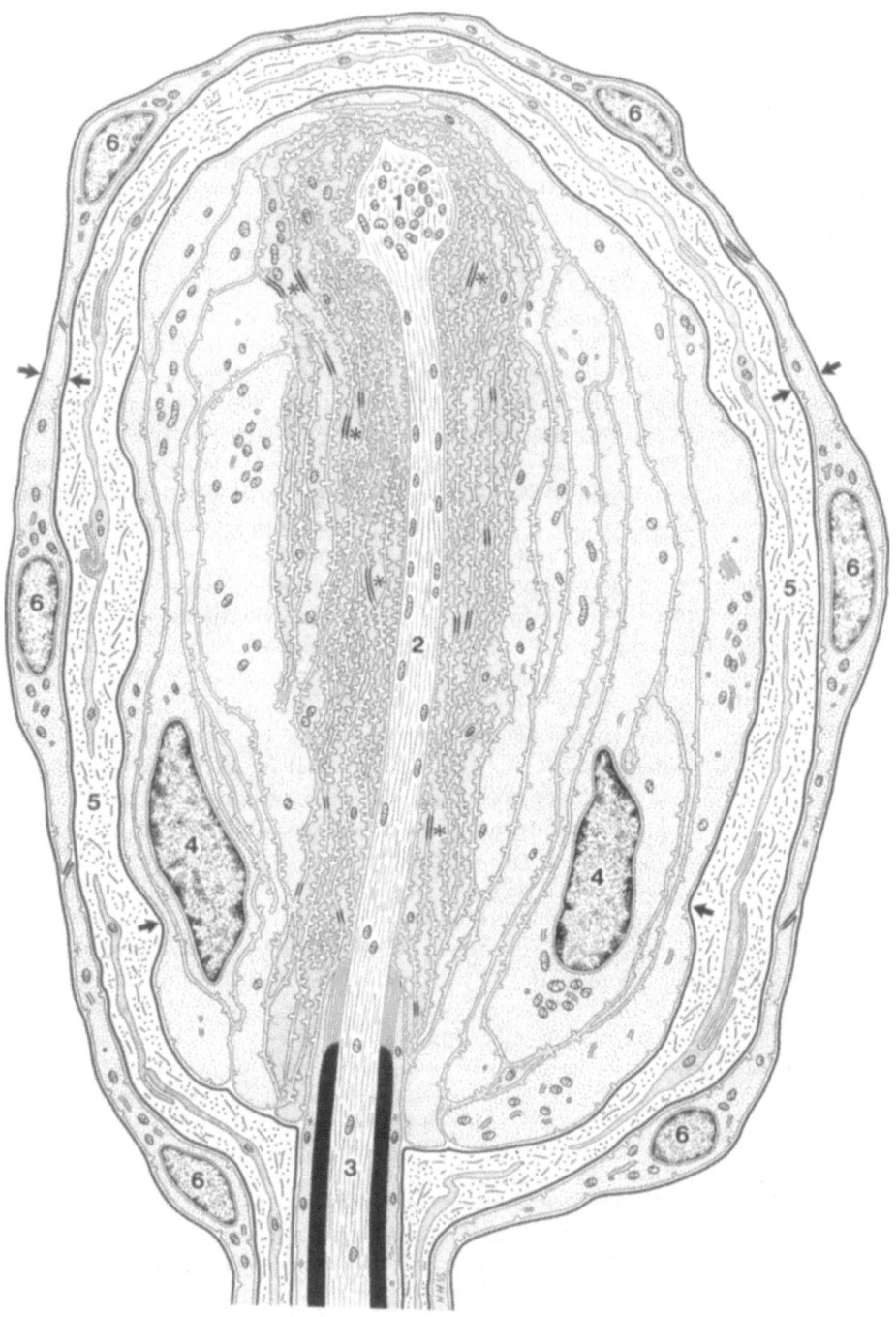

Fig. 5i

axolemma and the cytoplasmic membrane of the inner lamella of the inner core. This structure resembles a "tight junction" (Fig. 5b).

The inner core of the corpuscle consists of lamellae of modified Schwann cells. The number of lamellae depends on the size of the corpuscle; their arrangement varies between different animals. In general, we can distinguish between two types of inner core: a) inner core without longitudinal clefts, and b) inner core with two symmetrical longitudinal clefts (Fig. 5e). Occasionally, in the nasal dermis of the cat, we found corpuscles whose inner core contained only one longitudinal cleft. In these cases the lamellar cytoplasm was relatively transparent and these may have been immature corpuscles.

Inner core without longitudinal clefts are found in simple corpuscles in the nasal dermis and the feet of the mole (Fig. 5b). The inner core consists of two to four cells, whose flat lamellae are twisted round the axon and its terminal. The outer lamellae are longer and broader and possess mitochondria and rough endoplasmic reticulum. The broadest lamella bears the longish nucleus, the curve of which follows the curve of the inner core. These broader lamellae also contain filaments and free ribosomes. Deep inside the inner core, the lamellae become thinner. The innermost lamella, which is contiguous with the axon and its terminal, is the thinnest. Here we find ribosomes and filaments but rarely mitochondria. All the lamellae are rich in pinocytotic vesicles both inside and outside (Fig. 5c, f). The lamellae are separated by gaps of about 0.1 μ, containing collagen fibres. In some places these spaces are broken up by structures resembling desmosomes (Fig. 5b, c).

The lamellae of the inner core cells are often arranged symmetrically. The lamellae of one cell alternate with the lamellae of the opposite cell in concentric semi-circles. The inner core is invested with a basal lamina with separates the outermost lamella of the inner core from the subcapsular space. In the distal part, the lamellae of the inner core cover the last node of Ranvier (Fig. 5d). At this point the basal lamine of the inner core fuses with that of the Schwann cell.

Inner core with two symmetrical clefts (Fig. 5e, f) occur in the end-organs of the nasal skin of the cat, in the pig snout, and with the typical Pacinian corpuscles in the digital skin of the monkey. The difference lies in the size of the inner core.

Fig. 5f. Detail of a Vater-Pacini corpuscle from the glabrous digital skin of the rhesus monkey ($\times 13000$). The nerve fibre (1) is inside the inner core. The longitudinal gap in the inner core ($\uparrow$) extends to the nerve fibre. Lamelle-free areas of the nerve fibre and the lamellae of the inner core are invested with basal lamine

Fig. 5g. Detail of Vater-Pacini corpuscle from the digital skin of the rhesus monkey ($\times 13000$). The subcapsular space (1) contains a fibrocyte (2)

Fig. 5h. Detail of a Vater-Pacini corpuscle from the digital skin of the finger of a rhesus monkey ($\times 13000$). Layers of collagen fibres (1) lie between cell layers covered on both sides with basal laminaa (2)

Fig. 5i. Semi-diagrammatic representation of a simple corpuscle with inner core from the nose of the mole. 1 Bead-like expansion of the axon with digitate processes. The middle portion of the axon (2) runs inside the inner core. The afferent nerve fibre (3) is myelinated. The inner core is formed of a lamellar system of Schwann cells (4). The lamellae are linked by desmosom-like structures (*). The subcapsular space (5) contains fibrocytes and collagen fibres. The capsule (6) is a continuation of the perineurium and is lined and covered with a basal lamina ($\uparrow$)

The inner core is bisected by two symmetrical clefts. In the smallest corpuscles, in the nose of the cat and the snout of the pig, the lamellae number about 10 (Fig. 5c); in the largest Vater-Pacini corpuscles in the digital dermis of the monkey, there are about 60 (Fig. 5e). The outer lamellae are again wider and contain rough endoplasmic reticulum, mitochondria, free ribosomes and filaments (Fig. 5a). The nucleus is contained in one of the outer lamellae; it is flat and its curve follows that of the inner core. The inner lamellae are narrower and contain filaments and ribosomes (Fig. 5b, c, f). Mitochondria are rare and the lamella widens at the point where they do occur. These lamellae regularly contain pinocytotic vesicles at both the cytoplasmic membranes (Fig. 5c, f). In general, however, their number is smaller than in corpuscles whose inner core has no clefts.

The individual lamellae of the inner core are covered with basal laminae on their inner and outer surface. These spaces between the lamellae are accordingly contained by the outer basal laminae of the inner lamellae and the inner basal laminae of the outer lamellae. Collagen fibres run parallel to the longitudinal axis of the corpuscle.

In some corpuscles the axon forks the last node of Ranvier. The inner core then also ramifies so that the terminal part of the corpuscle contains two or more inner cores. All these inner cores have two symmetrical clefts.

The longitudinal cleft extend to the basal lamina of the axon and contains collagen fibres. We never found fibrocytes or their processes in these clefts. In some places one finds digitate processes of the axoplasm of the nerve fibre. The processes too are enveloped by a basal lamine.

The subcapsular space (Fig. 5e, g) varies in size. It is wider in large corpuscles than in smaller ones, and in those without longitudinal clefts, it is limited on the inside by the basal lamina of the inner core and on the outside by the basal lamina of the capsule. In corpuscles whose inner core has a symmetrical cleft, the latter communicates with the subcapsular spaca.

The capsule is perineural in origin. The width of the capsule depends on the size of the corpuscle. Its structure is the same for all corpuscles. The layers of the perineurium continue without interruption to form the layers of the corpuscular capsule. The capsule accordingly consists of stratified perineurial cells. The capsules of the small corpuscles in the nasal dermis of the mole consists of one to two layers, those of the cat and pig of one to five (Fig. 5a), whilst the largest, i.e. Vater-Pacini corpuscles, contain capsules with up to 30 layers of flat cells (Fig. 5e). The cytoplasm of the cells contains a rough endoplasmic reticulum, free ribosomes and mitochondria. Pinocytotic vesicles with a diameter of approximately 600 Å occur on both sides of the cytoplasmic membrane of the capsular cells. The cell nuclei are oval; in their vicinity we find the Golgi apparatus and filaments extending into the cytoplasm of the processes. The cells of each layer are linked by desmosomes (Fig. 5h).

The outer and inner surfaces of each layer are covered by a basal lamina. The basal laminae on the inner surface of the outer lamella and the outer surface of the inner lamella are separated by a gap 0.3–1.0 μ wide. The capsule accordingly shows a periodicity: basal lamina—inner cytoplasmic membrane— basal lamina—gap—basal lamina etc. (Fig. 5h). The spaces contain bundles of collagen fibres running parallel and perpendicular to the curve of the corpuscle. We found no fibrocytes in this space. The capsule frequently invests groups of 2–5 inner cores. The innermost layer of the capsule then forms a septum separating the inner cores.

In general it is true that the deeper a corpuscle lies in the dermis, the more layers there are in the capsule, and vice versa—the nearer a corpuscle is to the surface, the fewer layers there are in the capsule. Corpuscles situated immediately below the basal lamina of the epidermis may be partially devoid of a capsule at this point.

C. Morphological Classification of the Mechanoreceptors of the Skin

With the aid of the electron microscope, we can distinguish three basic types of nerve endings in the skin. The first type represents the epidermal nerve ending; the dermal nerve endings belong to the second or third type.

Type I. A characteristic feature of the *epidermal nerve endings* is the absence of the Schwann cell. These endings include the free nerve endings and the Merkel endings. The free nerve endings were definitely seen only in the mole. The nerve fibres end in a bead in the granular layer of the epidermis. The Merkel endings consist of a neural expansion in contact with a Merkel cell. They are found in the basal-cell layer of the epidermal cones and the glandular ridges.

Type II. The second type is represented by the *bulboid nerve endings* of the dermis. They consist of a nerve terminal invested with lamellae of Schwann cells. None of these bulboid nerve endings has a typical perineurial capsule. The nerve endings are situated in the stratum papillare of the dermis just below the basal membrane of the epidermis. The dendritic bulboid nerve endings, on the other hand, form complicated structures. They form palisade nerve endings around hairs, lanciform endings around sinus hairs and Meissner corpuscles in the digital skin of the monkey.

Type III. The third type of nerve ending, *encapsulated corpuscle with inner core*, differs from the second type in that the lamellae of the Schwann cells form an inner core. The inner core is invaginated into a capsule which is a direct continuation of the perineurium of the afferent nerve fibre and has the same structure as the perineurium. The size of the corpuscle depends on its position in the dermis. Those near the surface are smaller than those in the deeper layers. The largest Pacinian corpuscle from the digital skin of the monkey can even be prepared macroscopically.

Diagram 6 shows a morphological classification of the above findings.

D. Types of Glabrous Skin and Their Innervation

The second part of our report is intended to demonstrate the dependence of the innervation, and accordingly the receptor network, on the form of the epidermal surface, the lower surface of the basal-cell layer of the epidermis and its counterpart in the stratum papillare. We were able to distinguish two main types in the glabrous skin of the animals examined by us: cone skin (nose of mole and cat) and ridged skin (digital skin of rhesus monkey).

1. The Cone Skin (Fig. 7 a, e)

The surface of the glabrous nasal skin of the cat and the mole is arranged in epidermal papillae of varying sizes. In the mole, the papillae are higher and wider

Endings	Structure	Localisation	Synonyms
Free nerve endings	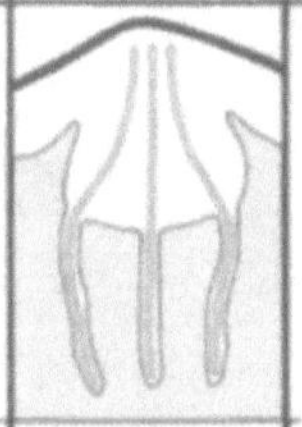	Epidermis: stratum granulosum	
----Type I----			
Merkel nerve endings	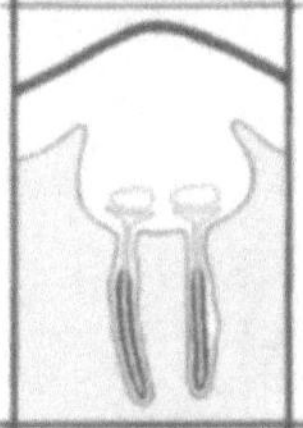	Epidermis: stratum basale	Brown bodies Iggo corpuscles Merkel cell-neurit complex
Simple bulboid nerve endings	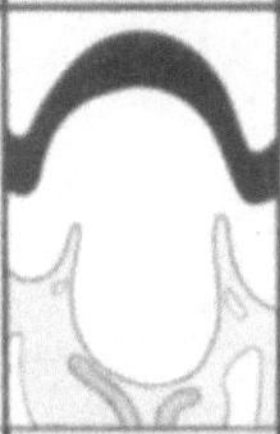	Dermis stratum papillare	Papillary nerve endings Free nerve endings of the dermis
----Type II----			
Dentritic bulboid nerve endings		Hair sinus hair: mesenchymal sheath, ridged skin: stratum papillare	Hair: palisade NE SH: lanciform NE RS: Meissner's corpuscles Ruffini end-bulbs Dogiel end-bulbs Genital corpuscles
Simple encapsulated corpuscle with inner core	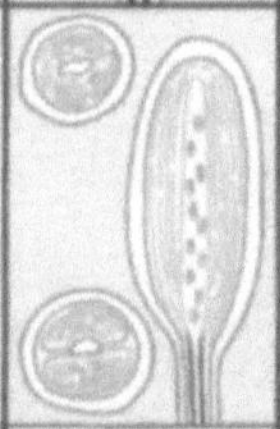	Dermis: below the epidermal cone	Krause end-bulbus Golgi-Mazzoni corpuscles Mucocutaneous end-organs Paciniform corpusles Innominate corpuscles
----Type III----			
Pacinian corpuscles		Deep layers of the dermis	Vater-Pacini corpuscles Rauber's end-organs

H. HESS

Fig. 6. Classification of the mechanoreceptors of the skin. Abbreviations: NE nerve ending, SH sinus hair, RS ridged skin

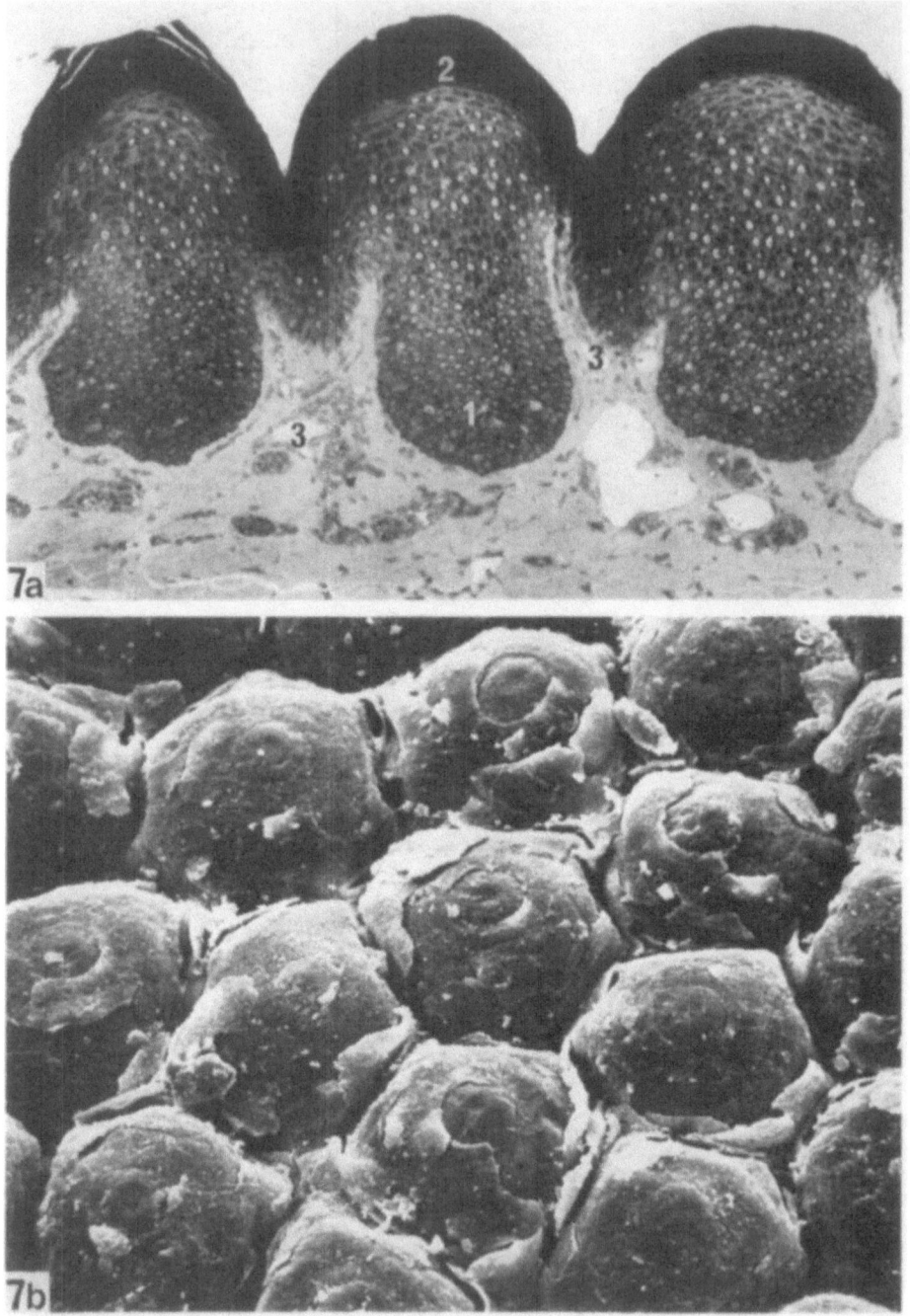

Fig. 7a Glabrous nasal skin of the cat. Semi-thin section perpendicular to the skin surface (×450). *1* Epidermal cone; *2* epidermal papilla; *3* capillaries

Fig. 7b Cones in the stratum basale of the epidermis of the glabrous nasal skin of the mole. Scanning electron microscope (×450)

Fig. 7c Detail of Fig. 7b (×40000). The surface of the basal-cell layer of the epidermis consists of folds. They correspond to the so-called rootlets

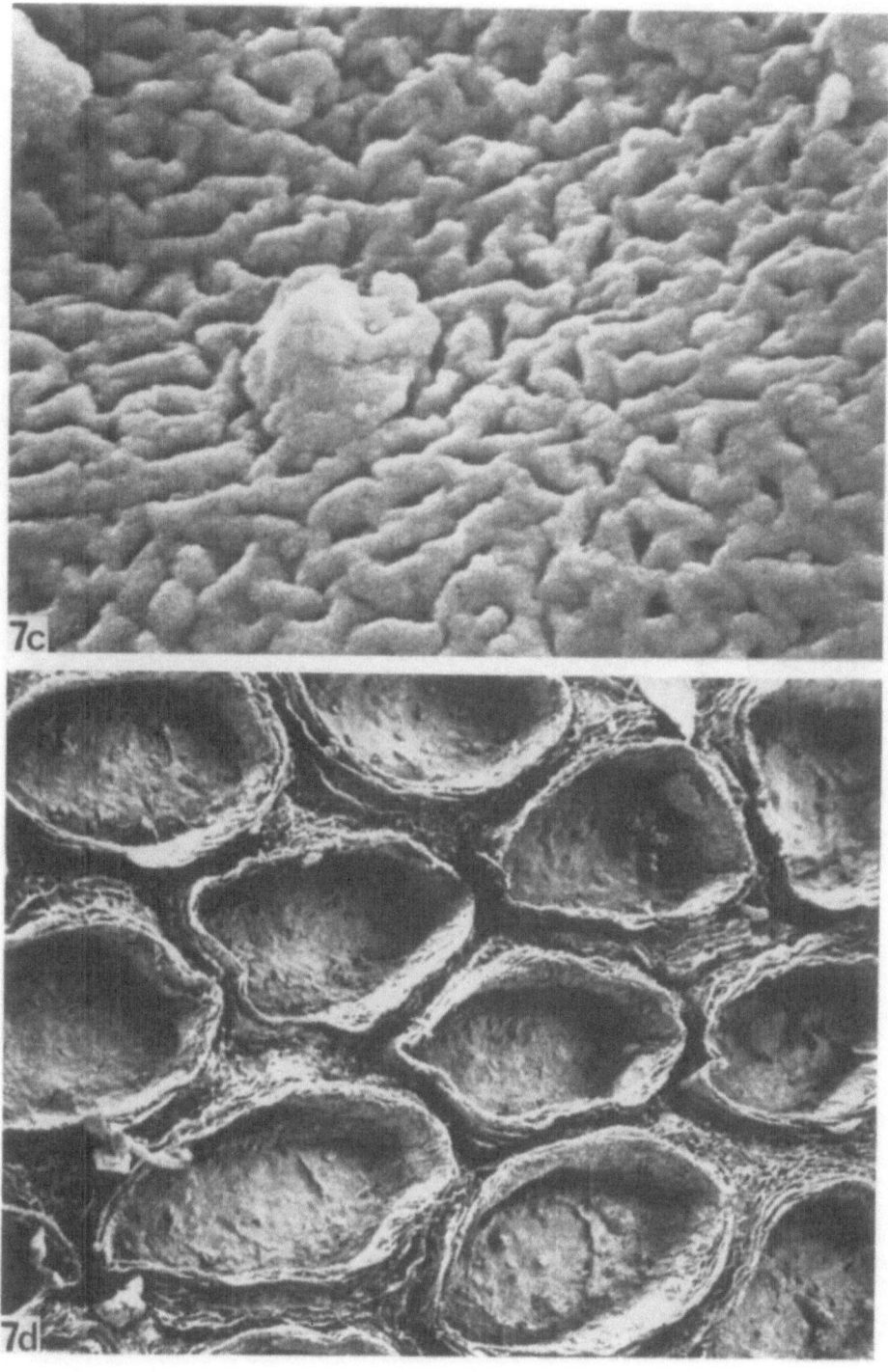

Fig. 7d The "papillae" of the stratum papillare of the dermis appear in scanning electron microscopy like the rims of basins. Epithelial papillae go down into the crypts. Blood capillaries occur between the structures ($\times 450$)

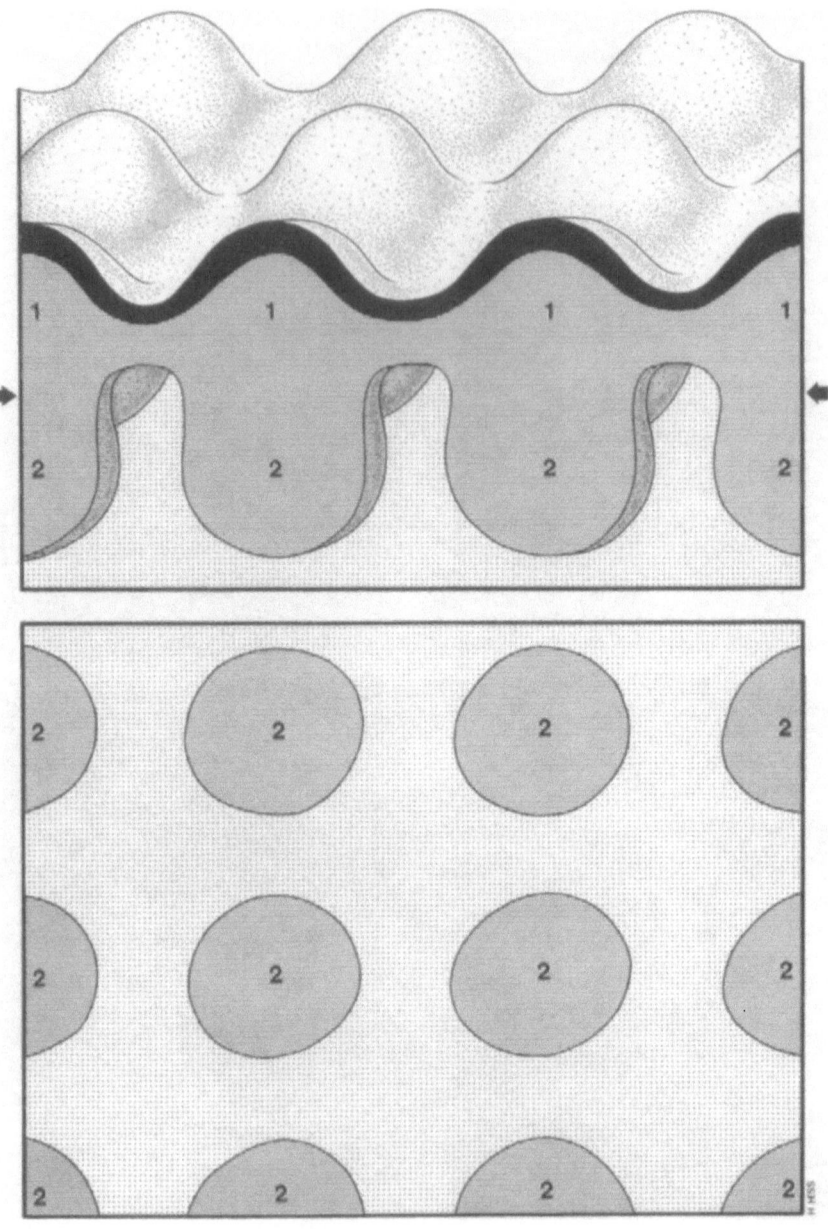

Fig. 7e Diagrammatic representation of the cone skin in the nose of the mole. The epidermal surface forms papillae (*1*) which continue as cones (*2*) in the dermis. The horizontal section (↑ plane) shows the regular arrangement of the cones in the dermis

in the region of the nostrils. Near the hairy skin of the nose, the papillae are flatter and smaller. Below, the inverted cones correspond to the surface papillae (Fig. 7b). They have a cushion-like widening at the base, the width being approximately 120 μ. The combined height of the epidermal papillae and the epidermal cones is approximately 200 μ. The epidermal papillae and corresponding cones are estimated to number approximately 5000.

The epidermal cones in the nose of the mole show central indentations (Fig. 7 b). This is where the afferent axons of the Merkel endings enter the epidermal cone. In the deeper layers, the Merkel cells are arranged around the central indentation. This, however, can be demonstrated only by light or electron microscopy (Fig. 2 a, b).

The lower surface of the basal-cell layer forms irregular folds (Fig. 7 c). Each fold is about 0.5 μ wide. In transmission electron microscopy, these folds appear as so-called rootlets. The basal lamina follows the folds.

The upper surface of the stratum papillare of the dermis mirrors the lower surface of the basal-cell layer of the epidermis. The dermal surface is not arranged in papillae but shows indentations in which the epidermal cones are invaginated (Fig. 7 a, d). The width and depth of the crypts correspond to the width and extend of the epidermal cones. The crypts are separated by connective tissue septa, which are shown by transmission electron microscopy to consist of a network of collagen fibres and fibrocytes. These septa are reached by processes from the blood sinus of the lower layer of the dermis. These processes combine to form a cavernous vascular bed, which isolates the epidermal cones from one another. Below the dermal crypts we find the corpuscles with an inner core, and deeper still, a nerve fasciculus consisting of myelinated and non-myelinated fibres. The myelinated fibres are situated in the centre, and the non-myelinated fibres at the periphery.

The arrangement of papillae and cones is less regular in the cone skin of the feline nose. The epidermal papillae fluctuate in height and width and so does the width of the base of the cones in the lower layers of the dermis.

Here again, the upper surface of the stratum papillare of the dermis mirrors the lower surface of the basal-cell layer of the epidermis; but in contrast with the nasal skin of the mole, the connective tissue septa separating the epidermal cones contain only capillaries. These form a capillary plexus.

The *innervation of the cone skin* is similar in the nose of the mole and the cat. The difference is confined to the number of nerve endings, which in turn depends on the thickness of the epidermis. The base of each cone contains Merkel endings, which number 3–5 in the mole (Fig. 11) and 5–15 in the cat. A simple corpuscle with an inner core is situated below each cone in the dermis, 1–2 in the nose of the mole and often up to 7 in the cat; these may be arranged in groups. A large number of free nerve endings in the epidermis is a typical feature in the cone skin of the mole, whilst free nerve endings are rare in the substantially wider and higher cones of the feline nose.

2. The Ridged Skin (Fig. 8 c)

We have found in the glabrous digital skin of the rhesus monkey that the surface and base of the epidermis are arranged in ridges, similar to that in man. On the lower surface, one can distinguish three different types of ridges: glandular ridges, adhesive ridges and cross ridges (Fig. 8 a, b). The glandular ridges are widest and deepest. They correspond to the surface relief of the skin ridges; every glandular ridge in the deeper layers continues into an epidermal surface ridge. The adhesive ridges run parallel to the former and correspond to the superficial

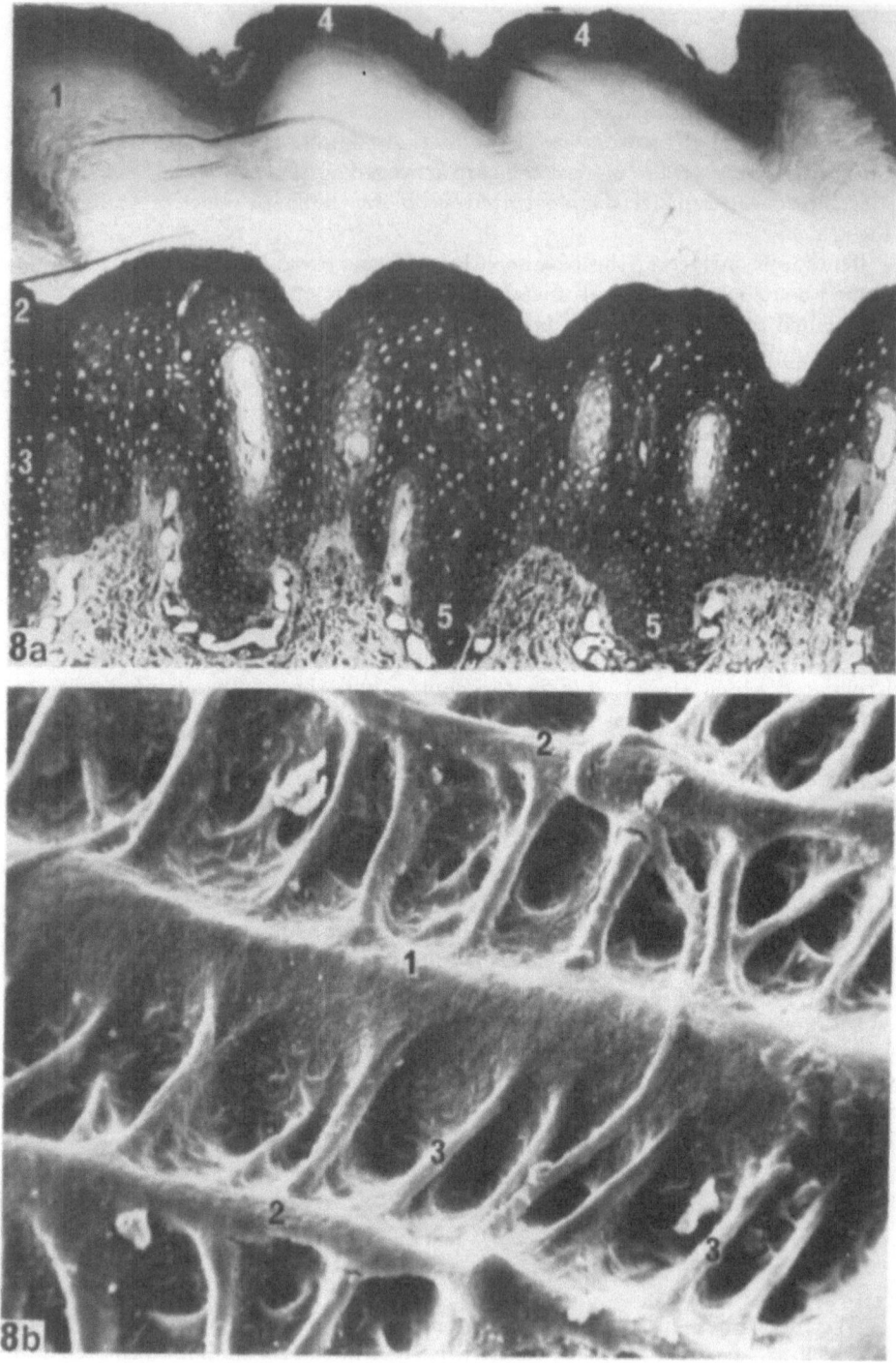

Fig. 8a In a section perpendicular to the skin surface (×240), the ridged skin of the finger tip of the monkey shows the layers of the epidermis: *1* horny layer, *2* stratum granulosum, *3* germinative zone. The epidermal surface is arranged in digital ridges (*4*). Each digital ridge correspond to a glandular ridge (*5*) underneath. The stratum papillare of the dermis contains a Meissner corpuscle (↑)

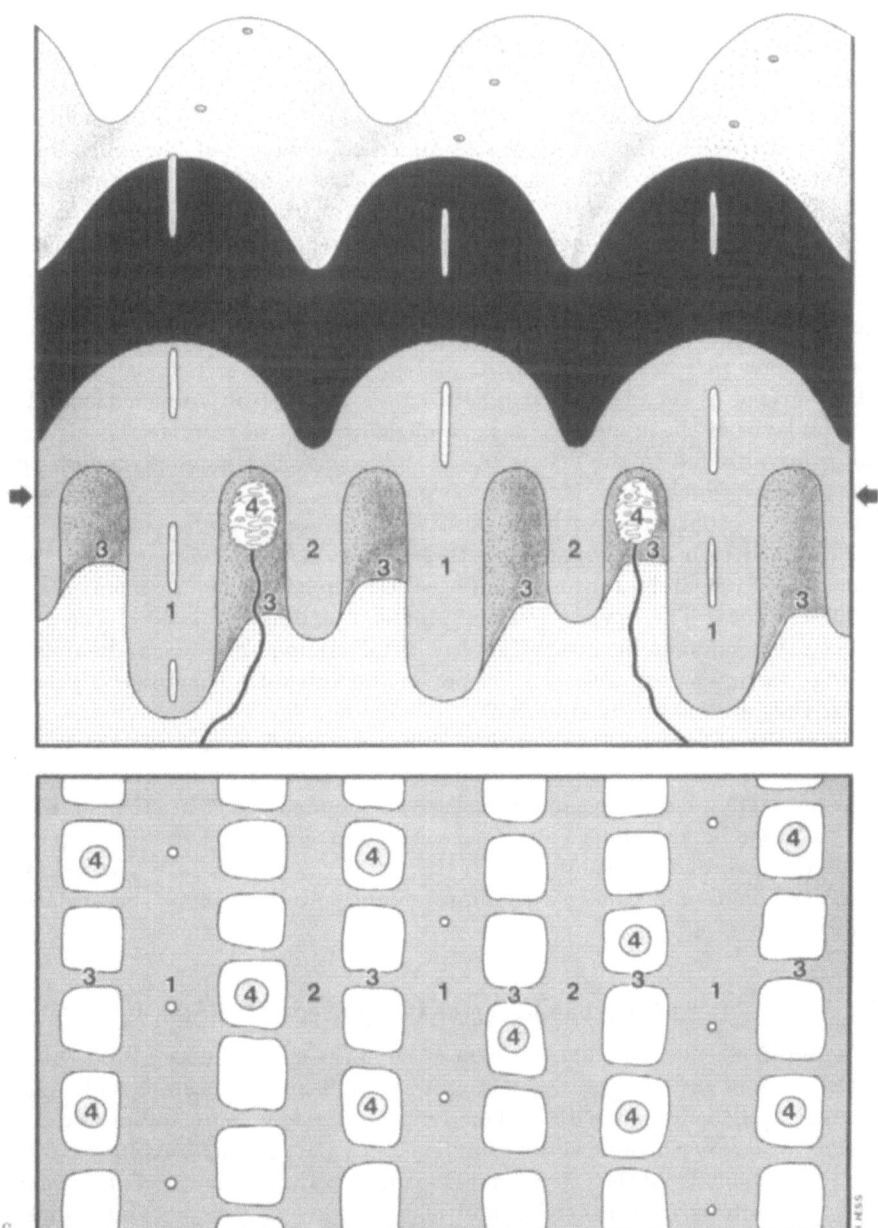

Fig. 8b Scanning electron microscopy of the basal-cell layer of the ridged skin (digital skin of the rhesus monkey; ×250). The surface is arranged in ridges. The narrower adhesive ridges (2) run parallel to the thicker glandular ridges (1) with the cross ridges perpendicular to both (3)

Fig. 8c Diagrammatic representation of the ridged skin of the finger of the rhesus monkey. Cross section: the epidermal surface is arranged in skin ridges which continue as glandular ridges (1) in the dermis. The grooves between the skin ridges correspond to the adhesive ridges (2) underneath. The cross ridges are perpendicular to both (3). The crypts formed in this way contain the Meissner endings (4). Horizontal section: (↑ cutting plane). The glandular and adhesive ridges run parallel (1 und 2) and the cross ridges (3) perpendicular to both. The crypts (light) are filled by connective tissue papillae of the dermis, containing Meissner endings (4) grooves between the epidermal ridges. The cross ridges run at right angles to the above two ridges. The total height of the glandular and epidermal ridges is approximately 2000 μ, and their width about 500 μ. The sweat gland ducts pierce these ridges. Cross sections through the ridged skin show that Merkel endings lie next to the entry points in the glandular ridges. The adhesive ridges are narrower and do not penetrate so deeply into the dermis (width 200 μ, total height of the adhesive ridge up to the groove of the surface relief, 1000 μ). The cross ridges are the lowest and narrowest.

The ridges divide the lower surface of the epidermis into crypts (Fig. 8 b, c). Each crypt is limited on one side by a glandular ridge, and an adhesive ridge, and by cross ridges perpendicular to both.

The surface of the stratum papillare again mirrors the lower surface of the basal-cell layer of the epidermis. It is arranged in rows of papillae. These dermal papillae are situated in the crypts of the epidermis. The papillae contain spiral nerve endings which form Meissner corpuscles. A particularly well developed capillary net is contiguous with the basal areas of the glandular ridges.

The ridged skin is innervated by three types of nerve endings. The Merkel endings (approximately 10) are situated where the sweat gland ducts enter the glandular ridges. The crypts contain Meissner endings (Fig. 4 b). The simple bulboid nerve endings lie just below the basal laminae of the glandular ridges. The large Pacinian corpuscles are deeper in the subcutis. Their size depends on how far they are below the epidermis.

Unlike the above two types of skin, which are hairless, the snout skin of the pig contains hairs which are separated from the dermal connective tissue by a blood sinus. They have the same structure as the sinus hairs in other mammals. The skin between the sinus hairs forms irregular cones and in this respect resembles the glabrous nasal skin of the cat. The Merkel endings are situated at the base of the epidermal cones, with simple corpuscles with inner core below the dermal cones (Fig. 2 a).

E. The Mechanoreceptor Complexes of the Skin

The form, size and number of the sensory nerve endings depends on the form and structure of the epidermis. One might say that all these components, i.e. the epidermis, and the dermis with its blood vessels and the nerve endings belonging to them, form a mechanoreceptor complex. Each part has its own function.

A tactile hair or vibrissa is a typical example of a mechanoreceptor complex (Fig. 9 k). It differs from other hairs in being thicker and longer, and originating deeper in the dermis. Because of the blood sinus which surrounds the hair shaft and hair bulb, the vibrissae are also known as sinus hairs.

1. Structure of the Sinus Hair[2]

Longitudinal section of a sinus hair, similar to that of ordinary hair, reveals different segments (Fig. 9 b, c): 1. Hair bulb and lower part of the hair follicle, 2. Hair follicle, 3. The section between the hair follicle and the duct opening of the sebaceous gland (Haarwulst), 4. The keratinised section above the duct of the sebaceous gland.

---

2 Nomenclature after Melaragno and Montagna (1953) and van Horn (1970).

The cross section (Fig. 9a, d) reveals two fundamentally different parts: the epidermal portion with the hair and the dermal portion with the blood sinus.

I. *The epidermal portion*, from inside towards the outside (Fig. 9d):

1. Hair: a) medulla, b) cortex, c) cuticle.

2. Inner root sheath: a) cuticle of the inner root sheath, b) Huxley's layer, c) Henle's layer.

3. Outer root sheath: a) inner cell layer, b) intermediate layer of polygonal cells, c) outer layer of columnar cells (basal-cell layer).

II. *The dermal part* from the inside towards the outside:

1. Basement membrane (Basal lamina).

2. Glassy membrane.

3. Mesenchymal sheath.

4. Blood sinus (ring sinus above, cavernous sinus below).

5. Capsule wall.

The thickening of the hair follicle is the result of proliferation of the polygonal cell layer.

Ordinary hair has the same structure. It is merely finer, the hair bulb is not so deeply embedded in the dermis, and the blood sinus is entirely absent.

2. Innervation of the Sinus Hair

The afferent axons form a nerve that enters the lower portion of the blood sinus, which at this point is cavernous. The nerve runs in a trabecula of the blood sinus and after a short distance ramifies into a number of myelinated fibres. These fibres supply all the three types of nerve endings found in sinus hairs: Merkel endings, dendritic bulboid nerve endings (lanciform endings) and corpuscles with inner core.

Merkel endings of the sinus hair are situated in the section of the hair follicle where they form a cuff (Fig. 9a, c). Their numbers is estimated to be between 400 and 800, depending on the size of the sinus hair. They are found in the basal-cell layer. They consist of a Merkel cell and a thickened axon. The Merkel cells are longish oval plates in squamous arrangement (Fig. 9c, d, f). They have many digitate processes which extend into the intercellular spaces of the adjoining cell layers. The processes frequently penetrate the basal lamina of the hair and thereby come in contact with the glassy membrane. The Merkel cells are linked to the surrounding cells by desmosomes. These are confined to the cell body and do not extend to the digitate processes. The cytoplasm of the Merkel cells has the same structure as that of the glabrous skin. The part of the cytoplasm lying opposite the nerve terminal again contains accumulations of osmiophilic granules (diameter 800–1000 Å).

The nerve terminal forms a disc (Fig. 9c, f); but unlike the Merkel endings in glabrous skin, it is in contact with the inner surface of the Merkel cell. The nerve disc contains mitochondria, vesicles, neurofilaments and neurotubules. The axolemma of the disc and the cytoplasmic membrane of the Merkel cell are thickened in some areas. The structure resembles a synapse, similar to the Merkel endings in hairless skin.

The nerve discs are dilatations of the axons which are myelinated in the mesenchymal sheath of the hair follicle. After losing its myelin sheath, the axon rises almost at right angles to the hair surface, enveloped by a Schwann cell,

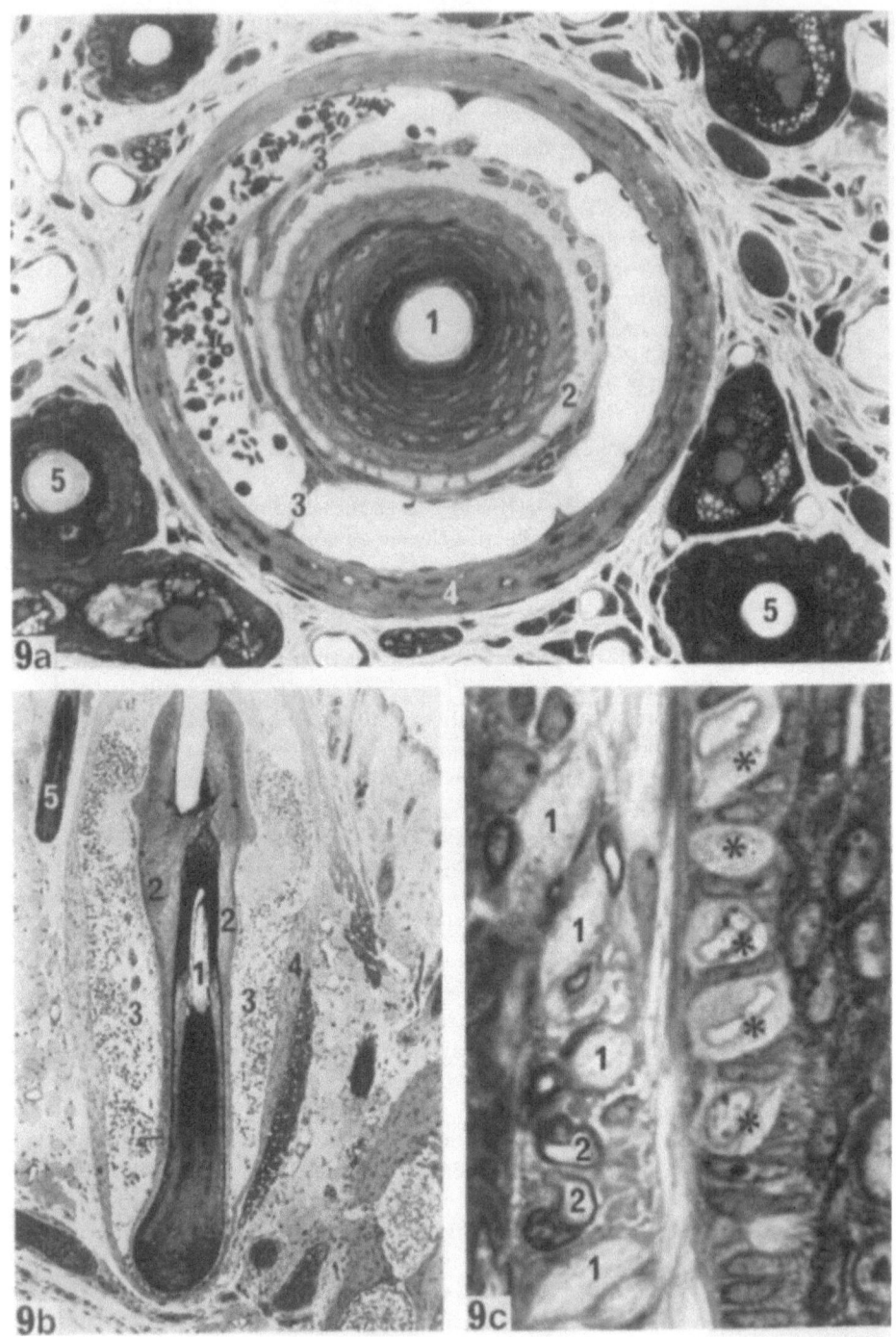

(Legend see p. 48)

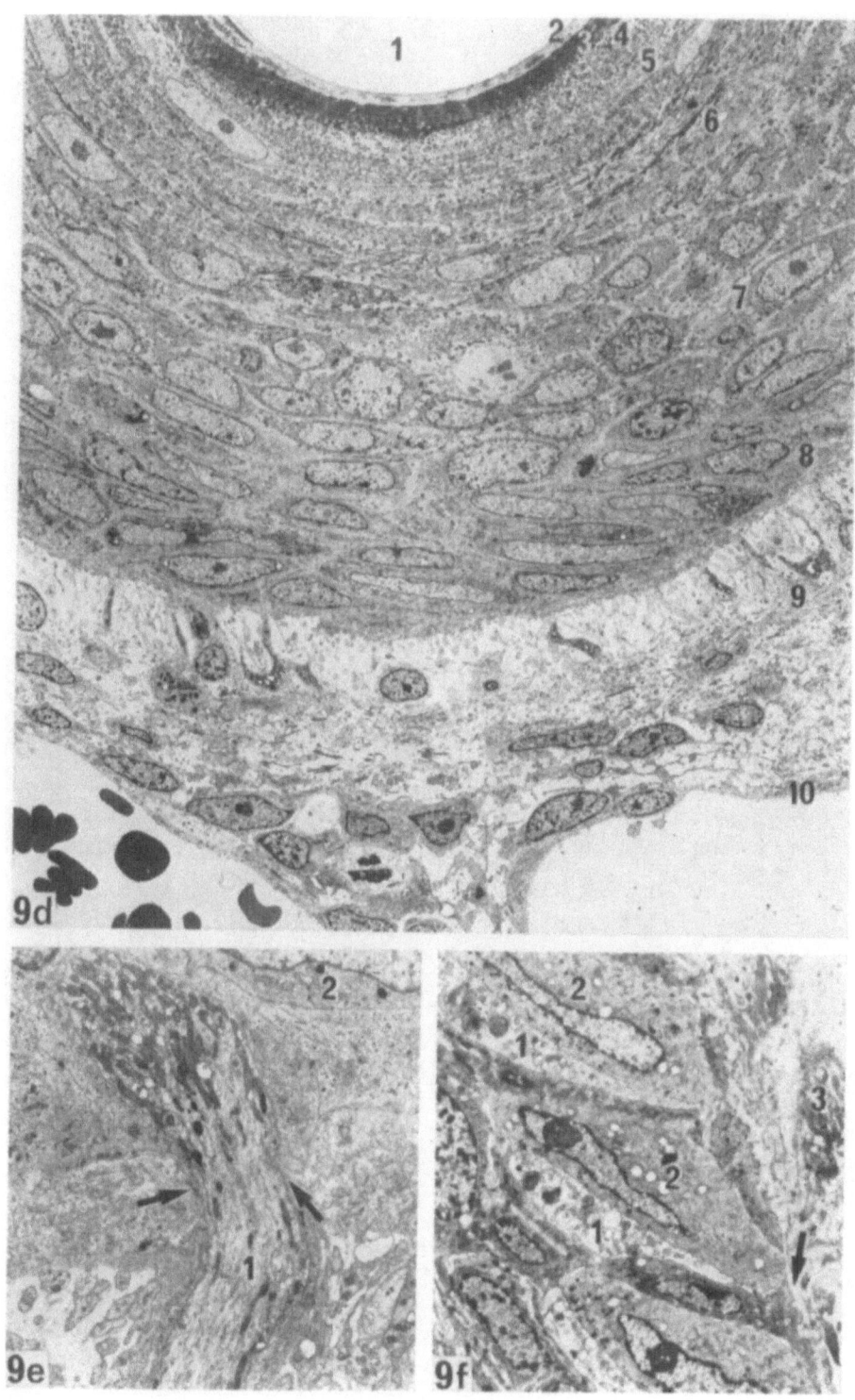

(Legend see p. 48)

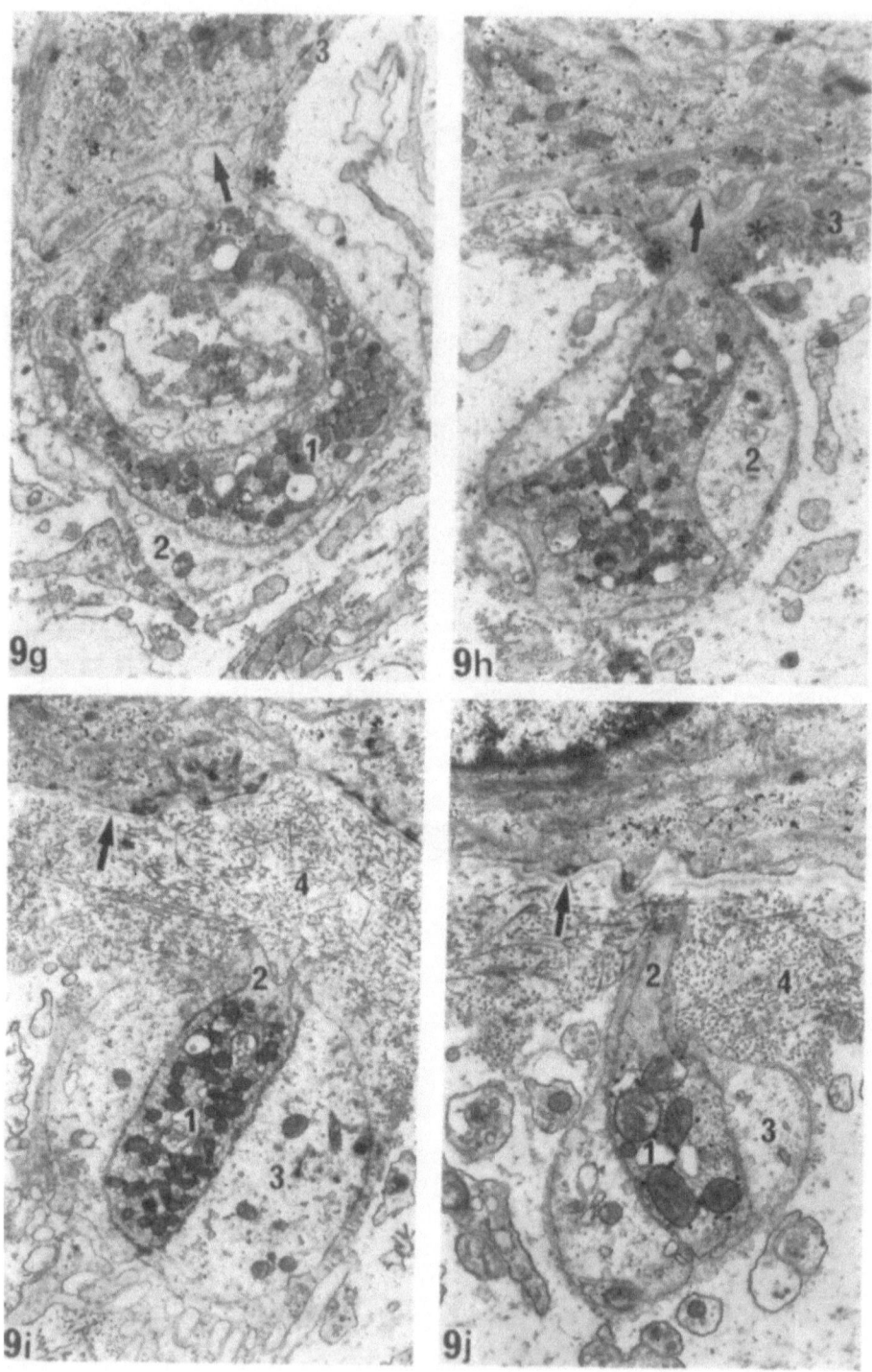

(Legend see p. 48)

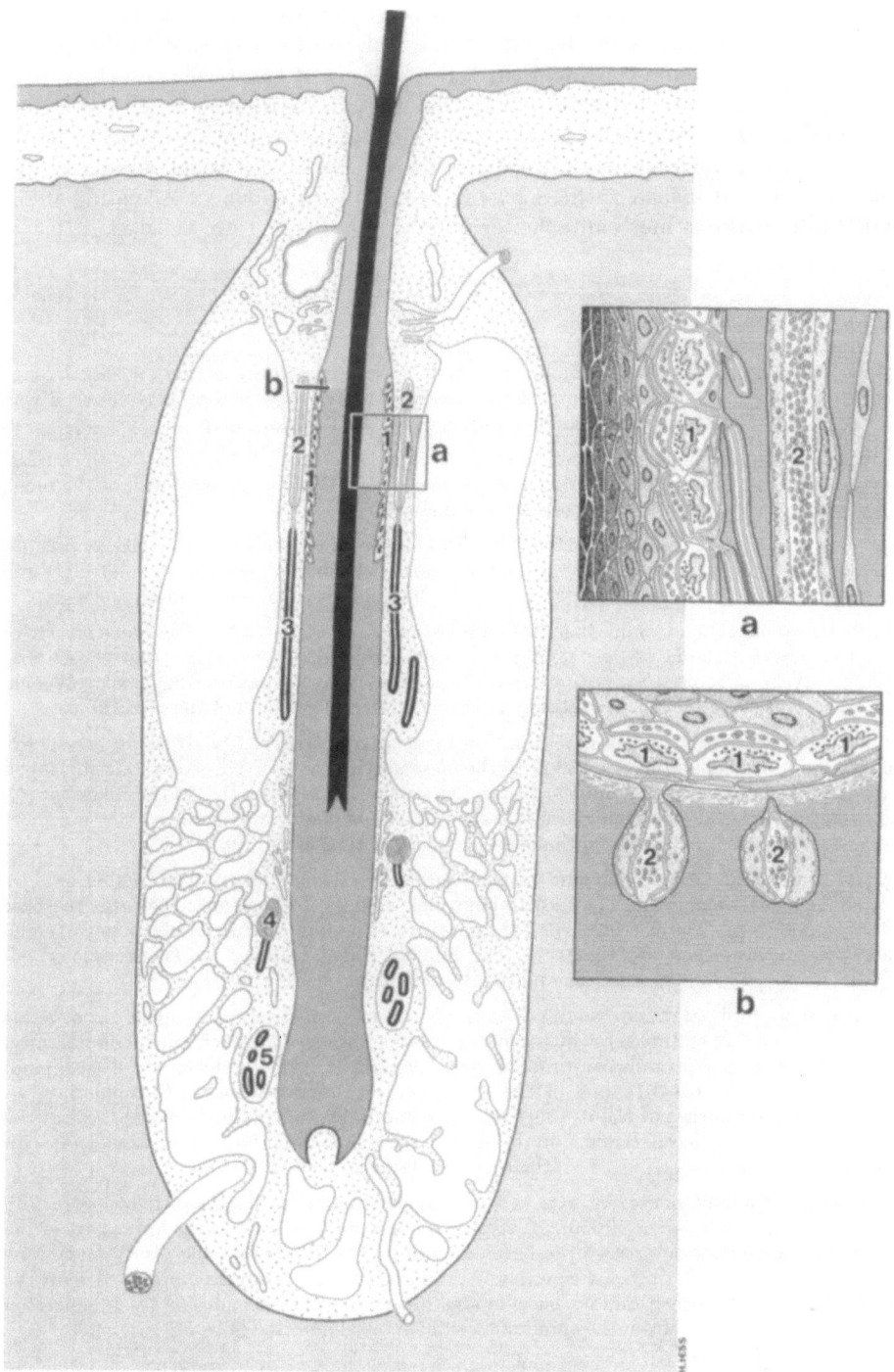

Fig. 9 k (Legend see p. 48)

reaches the basal lamina without a myelin sheath. After penetrating the glassy membrane, the basal lamina of the Schwann cell fuses with that of the hair (Fig. 9e). The axon continues through the layer of basal cells and there may ramify repeatedly. We estimate that in this way, the axon may reach more than eight Merkel cells. Compared with the lanciform endings, which are often in contact with the glassy membrane and sometimes also with the basal lamina of the sinus hair, the cytoplasm of the Schwann cell of a continuous axon, ending at the Merkel cell, contains more mitochondria and fewer pinocytotic vesicles.

Fig. 9a Sinus hair from the upper lip of a mole, in cross semithin section (×200). *1* Sinus hair; *2* mesenchymal sheath; *3* septa of the mesenchymal sheath in the blood sinus; *4* wall of the blood sinus; *5* ordinary hair next to sinus hair

Fig. 9b Longitudinal semi-thin section of sinus hair from upper lip of a cat (×70). *1* Hair; *2* layer of Merkel cells in the basal layer of the sinus hair; *3* blood sinus; *4* wall of the blood sinus; *5* ordinary hair

Fig. 9c Detail from Fig. 9b (×1200). The basal layer of the hair contains Merkel cells (*) in squamous arrangement. The mesenchymal sheath contains bulboid nerve endings (*1*) and myelinated nerve fibres (*2*) which supply the bulboid nerve endings and Merkel cells

Fig. 9d Cross section of sinus hair from upper lip of a cat (×1200). The hair consists of several layers: *1* Medulla of hair (torn); *2* hair cuticle; *3* and *4* cuticle of root sheath and Huxley's layer; *5* Henle's layer; *6* internal cell layer; *7* intermediate cell layer of polygonal cells; *8* basal-cell layer; *9* mesenchymal sheath; *10* endothelium of the blood sinus

Fig. 9e Sinus hair from feline upper lip. Continuation of the nerve fibre from the mesenchymal sheath into the epidermis (×4600). In the mesenchymal sheath, the axon (*1*) is enveloped by a Schwann cell. The basal lamina is of the Schwann cell fuses with that of the epidermis (↑). The axon is thickened in the epiderm and there contains many mitochondria. After a short distance, the bead ends at the Merkel cell (*2*)

Fig. 9f Sinus hair from pig snout in longitudinal section (×3000). The Merkel endings show squamous arrangement. The terminal expansion (discs) (*1*) are turned towards the base of the Merkel cells. The Merkel cells (*2*) contain osmiophilic granules, mitochondria, Golgi apparatus, ergastoplasm and filaments. The bulboid nerve ending (cut longitudinally) (*3*) protrudes at one point into the glassy membrane (↑)

Fig. 9g and h Bulboid nerve endings from the sinus hair from the upper lip of a cat (×10000). The nerve ending (*1*), which contains a large number of mitochondria, may ramify. It is enveloped in a Schwann cell and its basal lamina (*2*). The Schwann cell has pinocytotic vesicles. The basal lamina of the Schwann cell fuses with that of the epidermis (*). Occasionally a protrusion of the ending (↑) may come in contact with the basal layer of the sinus hair cells. The basal lamina of the sinus hair is linked by hemi-desmosomes (*3*) with the basal-cell layer

Fig. 9i and j Bulboid nerve endings in the mesenchymal sheath of the feline upper lip (Fig. 9i: ×10000; Fig. 9j: ×20000). The glassy membrane is thicker in the lower part of the Merkel cuff. A digitate process of the bulboid ending (*1*) extends into the glassy membrane of the sinus hair (*2*). The process contains neurofilaments. The endings is invested with the lamella of the Schwann cell and its basal lamina (*3*). The glassy membrane (*4*) is separated from the epidermal cell by a basal lamina (↑)

Fig. 9k Semi-diagrammatic representation of a sinus hair with three types of mechanoreceptor. The Merkel endings (*1*) are situated in the basal cell layer of the epidermis of the sinus hair. The dendritic bulboid nerve endings (*2*) are inside the mesenchymal sheath, the simple corpuscles with inner core (*4*) are in the lower part of the mesenchymal sheath. The afferent axons (*3* and *5*) of the mechanoreceptors are myelinated. (a) Detail: longitudinal section; (b) detail: cross section

Dendritic bulboid nerve endings (lanciform endings) are situated in the connective tissue of the mesenchymal sheath opposite the hair follicle, approximately at the same level as the cuff of Merkel endings (Fig. 9 a, c, d, g, h, i). The length of the lanciform endings is about 300–400 μ. On the basis of their relation to the glassy membrane, which is very thin in the upper region of the hair follicle but in the lower regions reaches a thickness of 3–5 μ, we can differentiate two sections of the lanciform endings—the upper section is in contact with the basal lamina of the hair, whilst the lower sections does not reach the basal lamina and ends at the glassy membrane.

The afferent axons of the dendritic bulboid nerve endings are situated in the mesenchymal sheath of the hair follicle and are myelinated. After losing its myelin sheath, the nerve fibre thickens and is invested with a modified Schwann cell.

The Schwann cell forms a simple lamella round the nerve ending. At the narrow end turned towards the hair, the lanciform endings forms digitate processes which are in contact either with the basal lamina (Fig. 9 g, h) or (in the lower region of the hair follicle) with the glassy membrane (Fig. 9 i, j). These processes are no longer invested with lamellae of the Schwann cells. They are covered only by a continuation of the basal lamina of the Schwann cell (Fig. 9 g–i). The lanciform endings may ramify, in which case one Schwann cell may contain two or more nerve endings (Fig. 9 g). The axoplasm of the endings contains many mitochondria, vesicles, neurofilaments, and neurotubules. The digitate processes contain only neurofilaments and vesicles.

The cytoplasmic membrane of the Schwann cells contains many pinocytotic vesicles, both on the inner and outer cell aspects. The cytoplasmic membrane of the Schwann cell is enveloped by a basal lamina. In cross section, these endings radiate from the sinus hair (Fig. 9 a, d).

Simple encapsulated corpuscles with inner core are also known as small Pacinian corpuscles. They are situated in the connective tissue of the mesenchymal sheath of the hair follicle in the distal third of the sinus hair. They are similar structure to corpuscles in hairless skin. They consist of an afferent nerve fibre which is myelinated outside the inner core. The inner core is formed by lamellae of the Schwann cell. The outer perineurial capsule consists of 1–2 layers of perineurial cells.

3. Innervation of Simple Hair

In order to obtain a complete picture, we also examined the innervation of normal hair (Fig. 10 a–c). All the hairs examined had only one type of nerve ending. The afferent axons of these endings are myelinated in the dermis. They ramify repeatedly and after losing their myelin sheath, approach the hair. The nerve terminals have a similar structure to that of the lanciform endings of the sinus hairs (Fig. 10 a, b). They are smaller, and the inner surface of the ending is always in contact with the basal lamina of the hair. We estimate that the endings per hair total about 60. They have no digitate processes, and a Schwann cell invests the portion not in contact with the basal lamina of the hair. The cytoplasmic membrane of the Schwann cell is furnished with pinocytotic vesicles. The axoplasm of the nerve terminals contains mitochondria, vesicles, neurotubules and neurofilaments.

As in the case of the sinus hair, the components of the hairless skin and their relevant nerve endings may be regarded as a mechanoreceptor complex. Such a

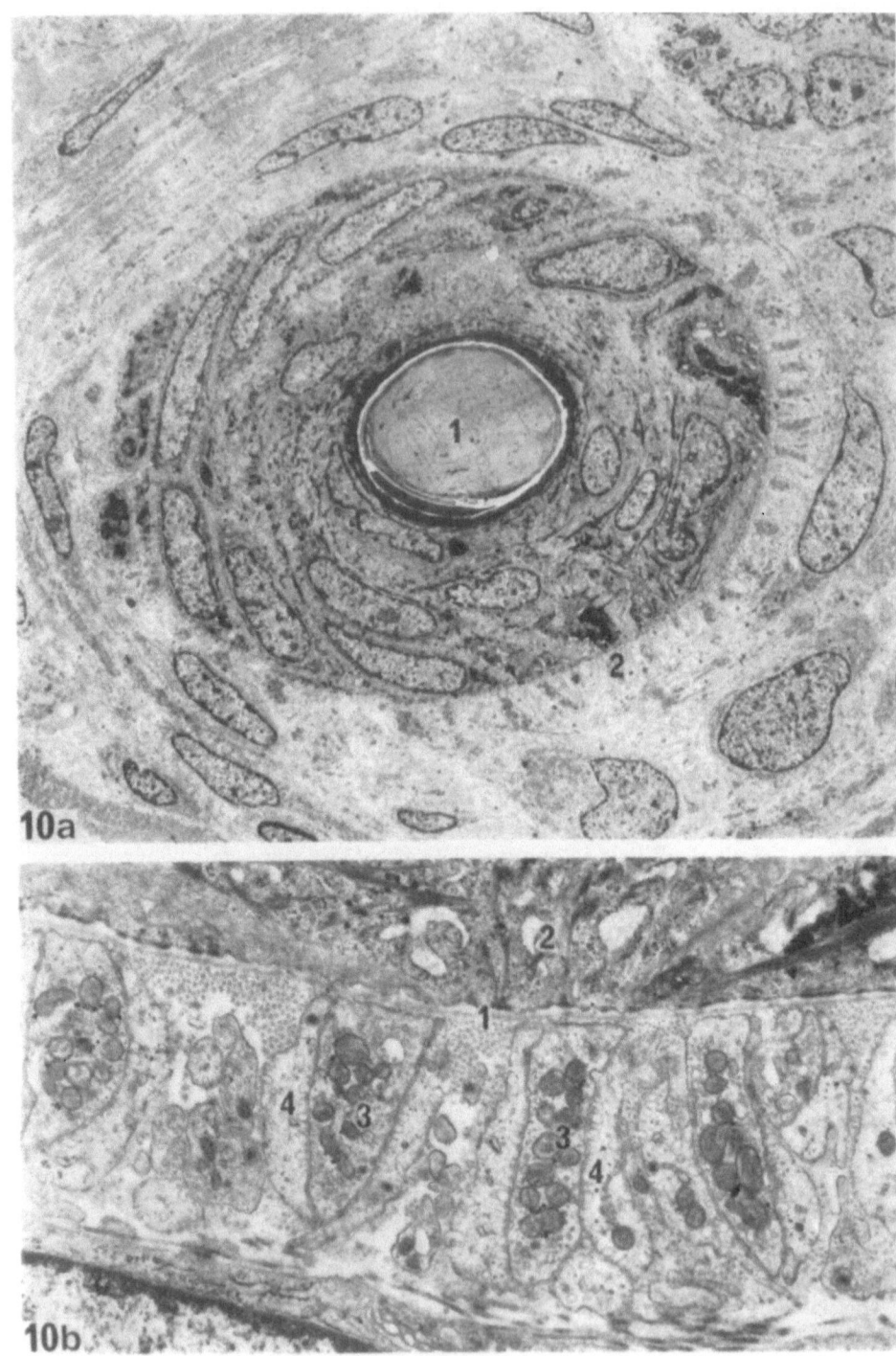

Fig. 10 a Cross section of a hair from the upper lip of a cat (×1100). A palisade of bulboid nerve endings lies close to the basal lamina of the hair. *1* Hair; *2* bulboid nerve endings

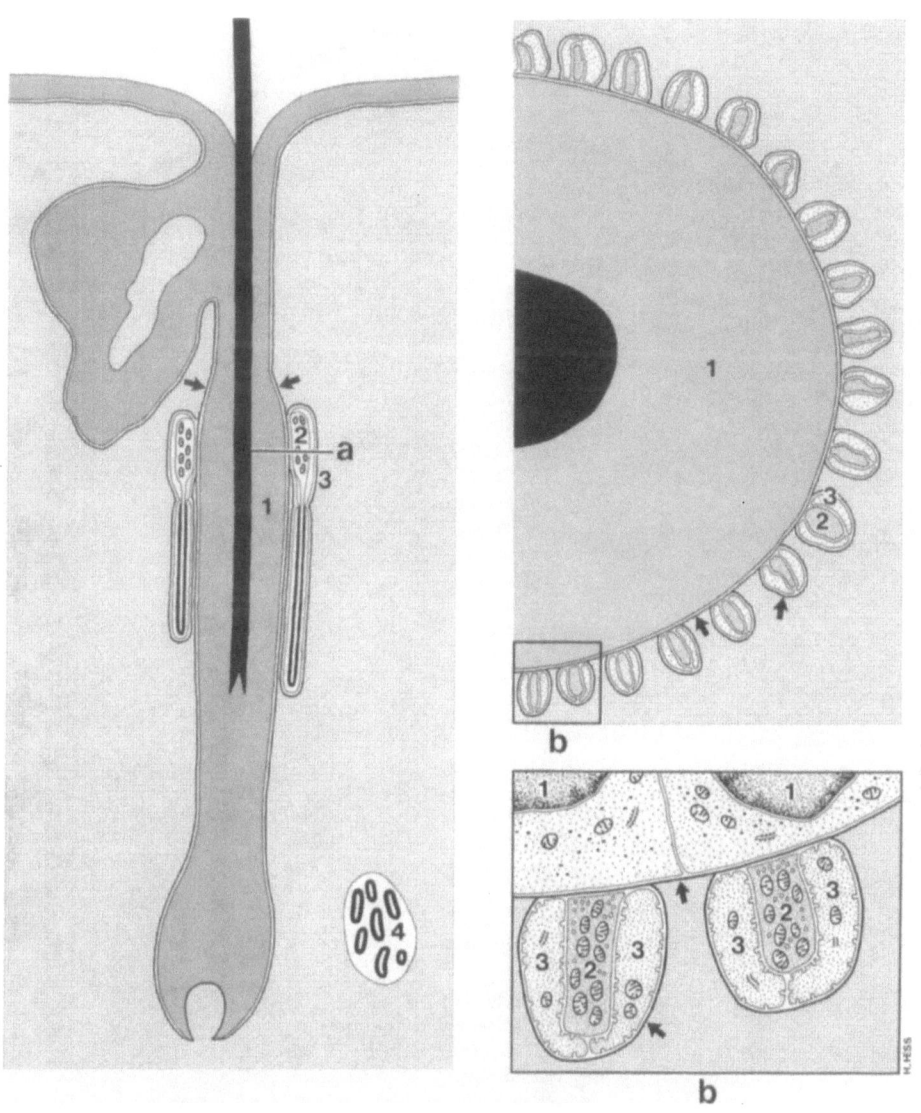

Fig. 10 b Cross section of a normal hair from the upper lip of a cat (×15000). A basal lamina (*1*) separates the outer layer of the hair (*2*) from the connective tissue of the hair follicle. The bulboid nerve endings (*3*) lie close to the basal lamina of the hair. They are enveloped by a lamellae of the Schwann cell and its basal lamina (*4*). The lamella of the Schwann cell contains pinocytotic vesicles. The bulboid nerve endings are naked at the point of contact with the basal lamina of the hair

Fig. 10 c Innervation of normal hair. The epidermal portion of the hair (*1*) is invested with a basal lamina (↑). The dendritic bulboid nerve endings (*2*) run parallel to the longitudinal axis of the hair and are in contact with its basal lamina. They are invested with a lamella of the Schwann cell (*3*) and its basal lamina (↑). (a) Cross section; (b) detail

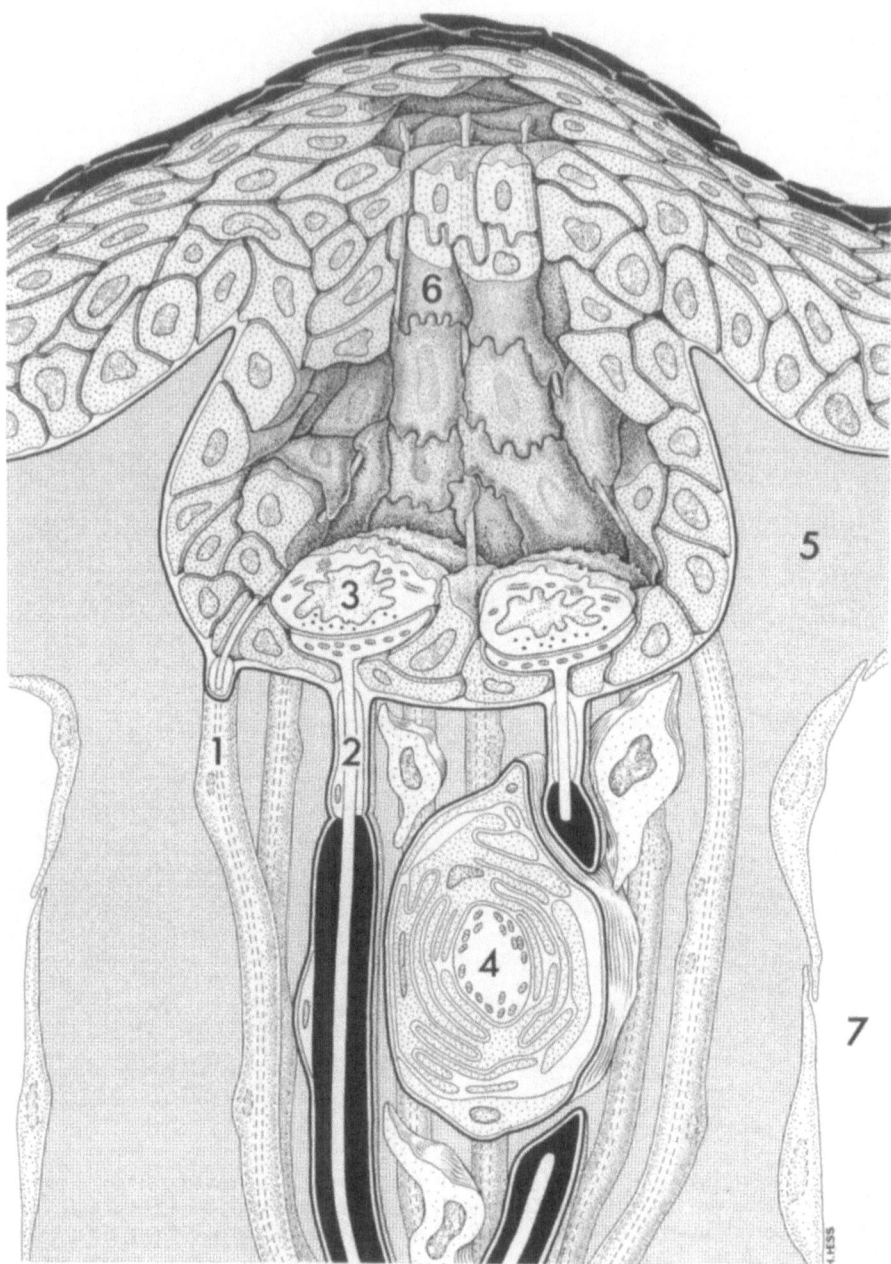

Fig. 11. Representation of a mechanoreceptor complex in the cone skin of the nose of the mole. *1* Afferent axon of the free epidermal nerve ending; *2* afferent axon of the Merkel ending with a nerve disc; *3* Merkel cell; *4* simple corpuscle with inner core; *5* stratum papillare of the dermis; *6* central column of the epidermal cell in a cone; *7* protrusion of the blood sinus. [Halata, Z.: Z. Zellforsch. **125**, 108–120 (1972), p. 116]

mechanoreceptor complex consists, in the cone skin (Fig. 11), of an epidermal cone with the corresponding epidermal papilla, the dermis surrounding the cone with its blood vessels, and the epidermal and dermal nerve endings belonging to them (free nerve endings, Merkel endings, and simple corpuscles with inner core). In the ridged skin, the mechanoreceptor complex may be regarded as consisting of the glandular ridges with the surface skin ridges, the stratum papillare of the dermis and the capillary network with the appropriate nerve endings of the epidermis (Merkel endings) and the dermal nerve endings (dendritic bulboid nerve endings = Meissner's corpuscles).

Discussion

A. The Epidermal Nerve Endings

1. Free Nerve Endings

The free nerve endings in the epidermis were first described by Langerhans (1868) in the human skin. The extensive morphological literature published since then on the subject reveals three basic problems: one, the variety of these nerve endings; two, the relationship of the nerve fibres and their endings to the cells of the epidermis and three, the layer of the epidermis in which the nerve fibres end.

Podcopaev (1869) found only one type of free nerve ending in the hairless skin of the rabbit. Tretjakoff (1902) described two types of free nerve endings in the snout of the pig. Botezat (1912), who studied the innervation of the hairy and hairless skin of different mammals (dog, cat, pig, mole and bat), even described seven types of free nerve endings. He differentiated these endings according to their diameter, the course of the axon and the number of intra-epidermal ramifications. According to Botezat (1912), the variability of these endings depends not only on the animal but also on the type of skin. For instance, the epidermis of the canine nose is more richly innervated than the nose of the bat, and the hairy skin of the dog has fewer nerves than that on the nose. Winkelmann (1960a) maintains that a thicker epidermis has more free nerve endings than a thinner one. It is, however, obvious from all the above studies that the actual terminal has the same structure, namely a bouton-like swelling. The variability relates primarily to the afferent axon, which may vary in its diameter and its ramifications. Our results are based on observation of these nerve endings in the nose of the mole. Electron microscopy shows these free epidermal nerve endings to be of similar structure. They resemble the endings described by Munger (1965) in the glabrous nasal skin of the opossum, and by Kadanoff (1971 a, b, c) in the glabrous skin of the mouth of the cow and the nose of the rat. The free nerve endings in the nose of the mole, however, show a regular and almost symmetrical arrangement, unlike those of the opossum, the cow and the rat. The axons, the number of which ranges between 20 and 25, run in each cone perpendicular to the skin surface, and all contain neurotubules, neurofilaments, mitochondria and vesicles. Unlike Munger (1965), we saw no degenerative processes in the terminal portions with boutons termineaux. We found no free nerve endings in the epidermis of the nasal skin of the dwarf pig or in the digital skin of the monkey. This contrasts with the results of Tretjakoff (1902) and Botezat (1912). However, these authors observed the endings under a light microscope, and their material was stained with methylen blue. In our view, this type of ending is so rare that it would be difficult to demonstrate it by electron microscopy. Hagen and Werner (1966a, b, 1967), and Hagen (1968) have described free nerve endings in the digital skin of the monkey and in the hairy and hairless skin of the dog. As these authors demonstrated free nerve endings only in the basal-cell layer of the epidermis, these may in fact have been the afferent axons to the Merkel endings.

The second problem concerns the relationship of the axon and its terminal with the epidermal cells. One group of authors (Eimer, 1871; Huss, 1900; Botezat, 1902a, b, 1903, 1907, 1908, 1912; Boeke, 1925, 1933a, b, 1940; Jaburek, 1927) maintains that part of the nerve fibres and all terminal expansions are intracellular. Another group of authors (Kadanoff, 1924b, 1928; Sasybin, 1930; Pieper, 1941; and again Kadanoff, 1971a, b, c), maintains that the nerve fibres and their endings are in extracellular form in the epidermis. Their reason for claiming an extracellular position for the axon may be due to the special relationship between the epidermal cells and the nerve fibres. Like Munger (1965), we came to the conclusion (Halata, 1972a) that the epidermal cells invest the axons as the Schwann cells do in the dermis. We may therefore say that the free axons do not run in the normal spaces between the cells but in tunnels formed by epidermal cells. In a substantial number of cases, a single epidermal cell may invest the entire circumference of an axon. In such cases we may refer to an invagination of the axon into the epidermal cell. Even in these cases, however, the axolemma of the nerve fibre and the cytoplasmic membrane of the epidermal cell are separated by an extracellular space that is 100–200 Å wide. We failed to observe a fusion of the axolemma with the cytoplasmic membrane.

The third question involves the layer of the epidermis that contains the boutons termineaux. Botezat (1912), Miller, Ralston and Kasahara (1958, 1960), and Quilliam (1966) found these endings in the granular layer. According to Cauna (1959), Munger (1965), and Ridley (1968), the free nerve endings even reached the horny layer where degeneration of the terminals was to be observed. Our findings are in complete agreement with those of Botezat (1912), Miller, Ralston and Kasahara (1958, 1960), and Quilliam (1966). In the nose of the mole, the terminals of the free nerve endings are found in the granular layer.

The function of the free nerve endings has not yet been fully clarified. Whilst earlier authors (Eimer, 1871; Botezat, 1912; Boeke, 1925) classified them as mechanoreceptors, Bielschowsky, as early as 1907, and later Quilliam (1966) claimed that the endings might be thermoreceptors. The question cannot be finally settled by morphological methods. At the same time, it seems to us that a mechanoreceptor function is suggested by the regular arrangement of the free nerve endings and their close connection with the keratinocytes in their palisade-like stacking in the epidermal cones of the nose of the mole since, as is the case with other mechanoreceptors of the skin, the nerve terminals are provided with auxilliary cells.

2. Nerve Endings with Specialised Tactile Cells (Merkel Endings)

The epidermal nerve endings with specialised tactile cells were described by Merkel (1875). They are confined to the germinative zone in both hairless and hairy skin. In the sinus hairs, or vibrissae, the endings are situated in the outer epithelial layer of the hair between the basal-cell layer and the polygonal cells. The fact that they occur in all mammals, in the hairless as well as in the hairy skin, has led to their being described in different terms by numerous authors. Merkel (1875, 1880) described the specialised cell as a touch cell and the complete structure of the cell and nerve disc as a touch receptor. Botezat (1912), Kadanoff

(1924 b), Boeke (1925), and Szymonowicz (1895, 1897) use the terms Merkel's touch cell and Merkel's touch corpuscle. Hoggan (1884) called the ending Brown's body, after his patroness. Miller, Ralston and Kasahara (1958, 1960) used the term dilated nerve ending, Cauna (1962, 1965, 1966) described it as Merkel's corpuscle, Sinclair (1967) as hederiform ending, and Munger (1965), and Munger, Pubols and Pubols (1971) as a Merkel cell-neurite complex. We used the term Merkel ending which has become accepted by morphologists as well as dermatologists, and which moreover is the one used in nearly all the textbooks on microscopic anatomy.

In the hairy skin too, this ending is know by different names. Pincus (1902, 1905) described it, if associated with hair, as a hair disc. The same term is used by Smith (1967, 1970), who studied these endings by electron microscopy. Straile (1960), and Mann and Straile (1965) apply the term "tylotrichic pad". Iggo (1963, 1966) and Iggo and Muir (1969), who studied these endings in the hairy skin of the cat, used the term "touch domes". Andres (1966a, b) calls it Merkel touch apparatus. Lipetz (1971) uses the term Iggo's corpuscle. These endings of the hairy skin and the sinus hair have the same structure as the Merkel endings in glabrous skin. We accordingly extended the term Merkel endings to them.

The Merkel ending consists of a Merkel touch cell and a nerve disc which is in contact with the Merkel touch cell. The Merkel cell has the same structure in all the animals examined by us. It is identical with the Merkel cells described, with the aid of the electron microscope, by a number of authors in other animals (Cauna, 1962, 1966, 1969; Kasprzak, Tapper and Craig, 1970; Munger, 1965; Quilliam, 1966; Andres, 1966a, b, 1969; Iggo and Muir, 1969; Patrizi and Munger, 1966; Munger, Pubols and Pubols, 1971), and in man (McGavran, 1964; Smith and Creech, 1965; Mustakallio and Kiistala, 1967; Breathnach and Robins, 1970; Breathnach, 1971a, b; Hashimoto, 1971, 1972). Merkel cells have been described not only in the hairy and hairless skin, but also in the gingiva of rats (Nikai, Rose and Cattoni, 1971), and in the oral mucosa of man (Hashimoto, 1972b).

The position of the nerve disc, on the other hand, does vary. In the epidermis of the hairless and the hairy skin, we find these terminal expansions in contact with the basal area of the Merkel cells, i.e. between the basal cell layer of the epidermis and the Merkel cells. In association with sinus hairs, however, the terminal bead is in contact with the inner surface of the Merkel cell, namely between the Merkel cell and the polygonal cells. We are inclined to interpret the variable position of the nerve disc in relation to the Merkel cell as an expression of a different mode of transmission of the mechanical stimulus to the Merkel cell and terminal expansion. In glabrous skin, the pressure is transmitted directly to the Merkel cell with the aid of the epidermal ridge or the epidermal cone. On the sinus hair, the cuff of Merkel cells is stimulated mechanically by bending the hair.

The number of Merkel endings in the glabrous skin depends on the structure of the epidermis. In the nasal skin of the mole, the cat and the dwarf pig, the Merkel endings are in the epidermal cone. The number of Merkel endings depends on the size and depth of the cone. The smaller cones in the nose of the mole, the total number of which Eimer (1871) estimates as approximately 5000, contain 3–5 Merkel endings. The cones in the nose of the cat have a wider base and contain 5–15 Merkel endings. In the pig, the cones penetrate deep into the epidermis and are higher than those in the other two animals, and they contain 20–30 Merkel endings. In the digital skin of the monkey, we find the Merkel endings only in the glandular ridges near the sweat gland ducts. The adhesive ridges, which correspond to the grooves of the skin surface, are devoid of Merkel endings. The arrangement of Merkel endings in groups, and their striking localisation in the cones, has induced some authors to describe this structure by a new name. Boeke, for instance (1925, 1932, 1933 b), described the epidermal cones in the nose of the mole as Eimer's organs, and Munger, Pubols and Pubols (1971) applied the term Merkel rete papilla to groups of 12 Merkel endings in the epidermal cones of the digital skin of the racoon. As these are purely quantitative differences, we feel that the introduction of new names is not only superfluous but also confusing for the classification of the receptors.

The Merkel cells extend a large number of digitate processes into the extracellular space between the cells of the stratum spinosum. In many cases they are invaginated in the cytoplasm of the keratinocytes. The digitate processes have no desmosomal links with the keratinocytes, in contrast to the cytoplasmic membrane of the cell body which is linked to the surrounding cells by desmosomes. Filaments extend from the cytoplasm of the Merkel cells into the digitate processes. The network of filaments of the Merkel cell is accordingly not confined to the cell body, but extends to its processes. By means of these processes, the Merkel cell is anchored in many epidermal cells. By this means one Merkel cell can receive mechanical stimuli from several keratinocytes in the area.

The cytoplasm of the Merkel cell contains osmiophilic granules. The latter are situated in the part of the Merkel cell cytoplasm that is in contact with the nerve disc. Opinions in the literature differ concerning the origin and chemical composition of the Merkel cell granule. According to Munger (1966a) and Hashimoto (1972a), they are PAS positive and contain gycoproteins (Munger, 1966a). Smith (1967, 1970), on the other hand, has observed a PAS negative reaction. Hashimoto (1972a, b) draws attention to the possibility of catecholamines ocurring in these granules, whilst Smith (1970) writes: "There is no evidence that they contain serotonin or catecholamines". McGavran (1964) discusses the similarity of the Merkel cells to the chromaffin cells of the adrenal medulla. The question remains unsolved.

The Merkel cell is invariably in contact with a discoid nerve terminal. Like Andres (1966a, 1969), we observed synapse-like structures in the region of contact between the Merkel cell and nerve disc (Halata, 1970, 1972a). Munger (1965), Smith (1967, 1970), and Munger, Pubols and Pubols (1971) and finally Munger (1971) failed to note a similar specialisation of the contact. Merkel (1875, 1880), as well as Botezat (1908), assumed that the mechanical stimulus was processed in the Merkel cell and transmitted to the nerve fibre expansion. Similar views are advanced by Andres (1966a, b). According to this concept, the Merkel cell receives the mechanical stimulus, "modifies" it, and transmits the processed information to the nerve fibre terminal.

From the physiological point of view (Iggo, 1963, 1966; Iggo and Muir, 1969; Munger, Pubols and Pubols, 1971), the Merkel ending is a slowly adapting mechanoreceptor (Type SA I). According to Schmidt (1971), it is a slowly adapting intensity detector with a mechanoreceptor function. Duclaux and Kenshalo (1972) maintain that its neural activity is affected by temperature. There is as yet no definite explanation of the specific function of the Merkel cell and of the mode of transmission of a mechanical stimulus from Merkel cell to the nerve disc.

B. Dermal Nerve Endings

The terminology and the subdivision of the dermal nerve endings were complicated by the fact that their occurrence may vary in different types of skin in the same animal, and in the same type of skin in different animals. These differences are not confined to quantity but extend to the special structures of these endings. This fact has persuaded many authors of the 19th and the beginning of the 20th century to consider anything that might be regarded as a variation of a type, as a separate type. A typical example of this attitude is represented by one of Botezat's papers (1912), in which he describes 31 different types of dermal nerve endings in the hairless skin of various mammals. This result is largely attributable to the silver impregnation method, as this allows examination only of the outline and not the detailed structure.

Electron microscopy has made it possible to reduce the multitude of dermal receptors to two types of nerve ending in the mammalian dermis: 1. bulboid nerve endings, and 2. encapsulated nerve endings with inner core. The nerve endings of one type have the same ultrastructure. Each type has its variants.

1. Bulboid Nerve Endings

This type can again be subdivided into two groups: a) simple bulboid nerve endings and b) dendritic bulboid nerve endings.

a) Simple Bulboid Nerve Endings

In order to distinguish between simple bulboid nerve endings [free nerve endings of Botezat (1912) and Cauna (1966)] in the stratum papillare of the glabrous skin and the continuing non-myelinated axons ending in the epidermis, it was necessary to establish definite criteria for a nerve terminal. Characteristic features of nerve terminal are a) the terminal expansion of the axon, b) the accumulation of mitochondria and vesicles, and c) investment with a Schwann cell which at this point is drawn out into a thin lamella, and the cell membrane of which is studded with numerous pinocytotic vesicles.

The afferent axons of the bulboid nerve endings of hairless skin come from the nerve plexus situated below the stratum papillare of the dermis. At this stage they are no longer myelinated. The nerve endings run parallel to the skin surface and are covered by a Schwann cell lamella and its basal lamina. In some areas the nerve terminal is enveloped only by the basal lamina (Cauna, 1966, 1969). It is situated immediately below the epidermis in the stratum papillare of the dermis of the nose of the cat, the pig's snout, and the ridged skin of the monkey. We failed to find it in the nasal dermis of the mole.

The function of the simple bulboid nerve endings has not yet been fully clarified. According to Cauna (1966, 1969) they may be mechanoreceptors. Iggo (1969), Hensel (1969) and Hensel and Iggo (1971) attribute the responses to warmth and cold to nonmyelinated sensory axons. This might apply to the afferent axons of the simple bulboid endings in the dermis.

b) Dendritic Bulboid Nerve Endings

The dendritic bulboid nerve endings form structures resembling palisades, spirals or coils. In association with hair, they form the so-called palisade nerve endings, with sinus hair, lanciform nerve endings, and in the stratum papillare of the ridged skin of the monkey, Meissner corpuscles.

The ultrastructure is the same in all these endings. The afferent axons are myelinated. After losing its myelin sheath, the axon ramifies. The terminal swellings are straight in the palisade-form and lanciform endings but form spirals in the Meissner endings.

Dendritic Bulboid Nerve Endings of the Hair and Sinus Hair

According to our estimation, the normal hair is supplied with approximately 15–20 myelinated nerve fibres. This corresponds approximately with the findings of Weddel (1945), who mentions 10–15 axons per hair. The diameter of these fibres varies between 4 and 6 μ. After losing its myelin sheath, the axon may ramify so that one axon may have 2–4 endings. One hair has approximately 60 bulboid nerve endings. These nerve endings have bulboid swellings. They run parallel to the longitudinal axis of the hair. The terminals contain accumulations of mitochondria and vesicles. The side facing the basal layer of the hair is flattened and in contact with the basal lamina of the hair. The outside of the endings is invested with a lamella of the Schwann cell and its basal lamina. The basal lamina of the Schwann cell may fuse with that of the hair. The lamella of the Schwann cell contains pinocytotic vesicles, a rough endoplasmic reticulum and mitochondria. Similar nerve endings have been described in human hair by Orfanos (1967) and Hashimoto (1972c), in the rat by Kadanoff, Seguchi and Villinger (1974), and in the mouse by Cauna (1969).

The dendritic bulboid nerve endings of the sinus hair are known in the literature as lanciform endings (Andres, 1966a, b; Munger, 1971). Their structure resembles that of the palisade nerve endings of normal hair. We estimate their number in the cat to be approximately 200. The afferent axon is also myelinated. The axon ramifies both before and after losing its myelin sheath. The terminals have bulboid expansions and run parallel to the longitudinal axis of the hair. They enveloped in a lamella of the Schwann cell and its basal lamina. Unlike the nerve endings of normal hair, the part of the ending facing the basal layer of the sinus hair forms digitate protrusions. These protrusions contain neurofilaments. In the upper third of the hair follicle, the glassy membrane is poorly developed. At this point the digitate processes may be in contact with the basal lamina of the sinus hair, or even come up directly against the basal cell layer of the sinus hair without any intervening basal lamina. In the lower two thirds, the glassy membrane is five times thick as in the upper third of the hair follicle. At this point the digitate processes do not reach the basal lamina but end in the glassy membrane.

According to Gottschaldt, Iggo and Young (1972), these endings are slowly adapting mechanoreceptors reacting to bending or turning (Gottschaldt, personal communication) of the sinus hair.

*Dendritic Bulboid Nerve Endings in the Ridged Skin of the Monkey
(Meissner Corpuscles)*

The bulboid nerve endings of hairless skin have been described by a number of authors in different types of skin. In the dermal papillae of the digital skin of monkeys and man, they are known as Wagner-Meissner or more usually Meissner corpuscles; in the penis of swine and horses, as spherical or cylindrical Krause end-bulbs (Ormea and Goglia, 1969), and in the human clitoris as genital corpuscles (Poláček and Malinovský, 1972). In the older literature, they are described as Dogiel's end-bulbs (Dogiel, 1893; Botezat, 1912), or as Ruffini's end-bulbs (Dogiel, 1903). We consider these endings to be variants of the same type differing only in their external shape, the latter being, in all probability, dependent on the skin surface. The regularly arranged ridged skin contains spiral Meissner endings (Cauna, 1956, 1959, 1966; Halata, 1972c). In the skin of the penis and clitoris, the dendritic bulboid nerve endings have a coil-like structure (Poláček and Malinovský, 1972).

The afferent axon of the Meissner ending is always myelinated. Our observations have shown that one ending may have 1–3 afferent myelinated axons. Cauna (1966) even describes 2–6 afferent axons in the human ridged skin. Their diameter is approx. 4–8 μ. After losing their myelin sheaths, the afferent axons may ramify repeatedly. The non-myelinated axons take a spiral course and end in a bulboid expansion. The expansions lie parallel to the skin surface. The Schwann cell invests the terminal with a lamella, covered on the outside by a basal lamina. The cellular membrane of the Schwann cell lamella is studded with pinocytotic vesicles, similar to the bulboid nerve endings of the sinus hair and ordinary hair. Between the terminals we find more Schwann cell lamellae which are reminiscent of irregularly arranged inner bulbs. The lamellae are interspersed with collagen fibres. The nuclei of the Schwann cells of the Meissner endings are on the outer sides.

In light microscope studies, these endings were described as having a capsule (Botezat, 1912), but electron microscopy has shown this structure to be composed of stratified connective tissue (collagen fibres and fibrocytes) and not a perineural capsule. Consequently the Meissner corpuscle would be more correctly described as a nerve ending than as a nerve corpuscle (Cauna, 1966; Quilliam, 1966; Munger, 1971). The term "corpuscle" should be reserved for endings with a typical perineurial capsule.

Since, morphologically speaking, the Meissner endings belong to the dendritic bulboid nerve endings, we believe that they are slowly adapting mechanoreceptors like the lanciform endings of the sinus hairs. A detailed physiological examination of the Meissner endings has yet to be carried out. It is generally assumed, however, that Meissner endings are mechanoreceptors. Munger (1971) suggests a rapid adaptation.

2. Encapsulated Corpuscles with Inner Core

On the basis of size and position, these endings are divided into small corpuscles in the stratum papillare of the dermis, and the large corpuscles in the lower layers of the dermis. The largest corpuscles which are situated in the fat deposits between the dermis and the muscular fascia, are generally known as Vater-Pacini corpuscles.

The small corpuscles of the stratum papillare of the dermis show great variability. Consequently they too are known in the literature by a variety names. Botezat (1903) and Boeke (1933b) called them "small Vater-Pacini corpuscles". Other authors have described them as Golgi-Mazzoni corpuscles (Poláček, 1966), muco-cutaneous organs (Winkelmann, 1960a, b), innominate corpuscles (Quilliam and Armstrong, 1963; Quilliam, 1966; Loo and Kanagasuntheram, 1972), paciniform corpuscles (Malinovský, 1966a, b, c) and simple encapsulated corpuscles (Poláček, 1966; Poláček and Halata, 1970).

The larger endings were described as Vater-Pacini corpuscles, or as Pacinian corpuscles. They are found not only in the lower layers of the dermis but also in the adventitia of various internal organs, e.g. in the urinary bladder (Shehata, 1970, 1972), in the peritoneum of the cat (Pease and Quilliam, 1957; Poláček and Mazanec, 1965), and in the periosteum of the middle ear (Gussen, 1970).

The presence of a capsule is a characteristic feature of this type. The capsule is a continuation of the perineurium of the axon (Shanthaveerappa and Bourne, 1963d; Poláček and Halata, 1965).

The structure of the perineurium has been studied by a number of authors, both with the aid of light microscopy (Shanthaveerappa and Bourne, 1962a, b, 1963a, b, c, 1964a, b) and electron microscopy. Structurally, the perineurium differs substantially from the endoneurium and the epineurium. The last two consist of collagen fibres and fibrocytes. The perineurium, on the other hand, consists of layers of flat cells alternating with layers of collagen fibres. The electron microscope reveals a layer of cells covered on the inside and outside by a basal lamina (Shanthaveerappa, Hope and Bourne, 1963; Thomas, 1963; Gamble and Eames, 1964; Waggener, Bunn and Beggs, 1965; Burkel, 1966, 1967; Cravioto, 1966; Waggener and Beggs, 1967; Ross and Reith, 1969; Kerjaschki and Stockinger, 1970; Halata, 1971b). The endothelium-like structure of the cellular layers is reflected in the various terms used to describe the perineurium. Lehmann (1957) uses the term Neurothel, Röhlich (1959) Perilemm, Clara and Özer (1959) stratum cellulare perineurii, Shanthaveerappa and Bourne (1962a) perineurial epithelium, and finally Cravioto (1966) perineurothel. We used the term perineurium (Poláček and Halata, 1965).

As the capsule of the corpuscle with an inner core has the same stratified structure as the perineurium of the nerves, we believe it to be a direct continuation of the perineurium.

The thickness of the capsule depends on the size of the corpuscle. The capsule of small corpuscles in the nose of the mole has 1–2 layers. The capsule of the large Vater-Pacini corpuscles in the cat's paw or the finger of the monkey has 20 to 30 layers.

The inner core of the corpuscle varies not only in size but also in structure. We differentiate inner cores with or without symmetrical longitudinal cleft. The inner core in the corpuscle in the nose of the mole, Quilliam's (1966) innominate corpuscle, has no longitudinal cleft. Unlike Quilliam (1966), however, we found longitudinally arranged collagen fibres in the interlamellar spaces. These fibres may have no relation to the axolemma of the nerve endings. The corpuscles in the nasal skin of the cat and the pig's snout have inner cores with two longitudinal clefts. These clefts were first described by Pease and Quilliam (1957), and Poláček and Mazanec (1965). These clefts, as well as the spaces between the lamellae of the inner core, contain collagen fibres. The function of these clefts has not yet been fully clarified. It may—as claimed by Pease and Quilliam (1957)—have a nutritive function. This view is supported by the fact that the lamellae of the inner core without cleft show greater pinocytotic activity than the corpuscles with two symmetrical clefts.

An interesting question is that of the origin of the inner core cells. The current view is that these cells are modified Schwann cells (Munger, 1966a, b; Andres,

1969; Poláček and Halata, 1970; Halata, 1972 b). This view receives support from the fact that the Schwann cell, like the cell of the inner core, has a basal lamina. The afferent axon is invariably invested with a myelin sheath of the Schwann cell. After the axon enters the inner core of the corpuscle, the basal lamina of the Schwann cell of the myelinated nerve fibre fused with that of the lamellar cells of the inner core. Similar findings have been reported in respect of the Herbst corpuscles in the bill integument of aquatic birds (Andres, 1969; Andersen and Nafstad, 1969; Halata, 1971 b). Nafstad (1972) advances the (speculative) suggestion that the cells of the inner core are derived from Merkel cells descending from the epidermis. This is rendered improbable by the fact that corpuscles with inner core are also found in mammalian articular capsules (Poláček, 1966), although no Merkel endings have been found in the articular capsules.

Occasionally, desmosom-like contacts are found between lamellae of an inner core (Andres, 1969; Halata, 1971 b, 1972 b).

The afferent axons of corpuscles with inner core are invariably myelinated. The last node of Ranvier contains the proximal portion of the inner core. The axon may fork at this point, in which case the inner core too will ramify. In cases of complete ramification, one afferent axon will be found with two inner cores in one capsule. This explains the arrangement of corpuscles in groups. In the nose of the mole, we observed a maximum of three nerve corpuscles in one group (Halata, 1972 b); in the nose of the cat, however, up to 8 corpuscles in one capsule. In these groups, the innermost layer of the perineural capsule forms a septum between the corpuscles. Similar arrangements of corpuscles are also found in the dermis of the pig's snout. In the older literature, these groups of corpuscles are described as Golgi-Mazzoni bodies (Dogiel, 1903; Botezat, 1912). We failed to see the "rosary-like corpuscles", in which two inner cores are threaded on one terminal, one after the other, like a rosary (Dogiel, 1903; Poláček, 1966). The axon terminal inside the corpuscle showed bead-like expansion. Similar expansions have been observed in the Vater-Pacini corpuscles in the mesentery of the cat (Poláček and Mazanec, 1965), in the Herbst corpuscles of birds (Quilliam, 1966; Munger, 1966; Halata, 1971 b), in the corpuscles of the feline nose (Poláček and Halata, 1970), and in the nose of the mole (Halata, 1972 b, c). The expansion extends digitate processes between the lamellae and into the longitudinal cleft of the inner core. We believe that these processes enable the terminal to register the slightest distortion in the structure of the inner core.

The over-all size of the corpuscles depends on how deeply it is situated below the epidermis (Poláček, 1968). The nearer the surface, the smaller the corpuscle; and vice versa, the deeper their position, the larger they are. The number of inner core lamellae increases with the size of the corpuscles, as to the layers of the capsule.

The smaller nerve endings have been studied, with the aid of the electron microscope, in the stratum papillare of the dermis of the following mammals: cow (Walter, 1962; Walter and Hebel, 1966; Hebel and Schweiger, 1967), cat (Andres, 1966 a, b, 1969; Ormea and Goglia, 1969; Poláček, 1969; Poláček and Halata, 1970; Spassova, 1970, 1973), tupaia (Andres, 1969; Loo and Kanagasuntheram, 1972).

The general rule is that as the nerve endings increase in size, their absolute number decreases (Poláček, 1968). The larger corpuscles in the deeper layers are less common than those near the surface.

Between the nerve terminal and the innermost lamella of the inner core, we observed contact with thickened portions of the opposite membranes. Similar contacts have been described by Munger (1966), Andres (1969), and Halata (1971 b, 1972 b).

We apply the term subcapsular space to the space between the capsule and the inner core [by Poláček (1966) "boundary space"]. In the avian Herbst corpuscles, this space is very large (Halata, 1971 b). In the mammals studied by us, the subcapsular space is larger in the Vater-Pacinian corpuscles than in the smaller corpuscles near the surface. The arrangement of the collagen fibres and fibrocytes found in this space is reminiscent of the endoneurium (Halata, 1971 b).

The problem of the function of the Pacinian corpuscles has been solved. Quilliam and Armstrong (1963), Quilliam (1966), Loewenstein (1971), and Schmidt (1971) have confirmed that the Vater-Pacini corpuscles are specialised for the reception of vibration stimuli. They belong to the rapidly adapting mechanoreceptors.

The smaller corpuscles with inner core are rapidly adapting receptors. Schmidt (1971) describes them as velocity detectors. This type includes, in all probability, the corpuscles with inner core in sinus hair which have been described by Andres (1966 a, b).

C. Morphological Classification of the Mechanoreceptors in the Skin

The sensory nerve endings can be subdivided according to various criteria. Lavrentjev (1947), for instance, differentiates mechanoreceptors, proprioreceptors, tensioreceptors, pressoreceptors, and chemoreceptors. Another possibility is to describe the receptors according to the organs in which they occur. Ivanov (1949, 1951) accordingly differentiates tendoreceptors, periostoreceptors, peritoneoreceptors, skin receptors, and fascioreceptors. In the traditional nomenclature, these endings generally bear the names of the authors who first described them. This nomenclature is based on light microscopy. The more recent light microscope studies of Poláček (1961, 1965, 1966), and Malinovský (1966 a, b, c) demonstrate that these nerve endings show a great variability of size and form. The large number of differences has led to variants of the same type being known by different names (Fig. 6—synonyms). A variant of one type ending is often confined to a single species. This caused many authors to describe variants of the same type by different names in different animals. Botezat (1912) provides an example of the complex classification of the nerve endings in hairless mammalian skin. He describes 38 types of nerve endings. In many cases the sole difference consists of a different number of afferent axons. Niculescu (1958) and Zabusov and Maslov (1961) have tried to simplify this complex classification. Another version is the division of the nerve endings into "free" and "non-free" nerve endings (Lavrentjev, 1947).

Poláček (1966) bases his classification on light microscope studies of the sensory nerve endings in the articular capsules of various mammals and birds. He divides the nerve endings into three groups according to their structure: free nerve endings, nerve endings with auxiliary cells and nerve endings with inner core. Our own classification, which is based on electron microscopy, results in a fundamentally similar division of nerve endings into three groups. Like Poláček (1966), we differentiate between free nerve endings and nerve endings having Schwann cells. Our principal criterion in sub-dividing nerve endings with Schwann cells is the presence of a capsule representing a direct continuation of the perineurium. Our sub-types are much smaller in number than those of Poláček (1966), the variety of which we attribute to the effect on their shape of the skin structure.

We have divided the mechanoreceptors associated with mammalian skin and its related structures into three types: Type I (epidermal nerve endings), Type II (bulboid nerve endings of the dermis), and Type III (encapsulate corpuscles with inner core).

Type I comprises the free nerve endings and the Merkel endings which are characterised by the absence of a Schwann cell. The nerve terminal is expanded and lies between the epidermal cells. In the Merkel endings, the terminal axon forms a disc which is in contact with a Merkel cell. In the free nerve endings, the afferent axon has no myelin sheath, whereas in the Merkel endings it invariably has a myelin sheath.

Type II nerve endings occur in the dermis close to the epidermis. These endings have a bulboid expansion which is invariably invested with a lamella of the Schwann cell and its basal lamina. In the simple bulboid nerve endings, the afferent axon is not myelinated but the dendritic bulboid endings invariably have a myelin sheath. On the basis of their ultrastructure, the Meissner endings of the ridged skin, and the lanciform and palisade-form nerve endings of the sinus hair and ordinary hair may be included in this group.

Type III covers the encapsulated corpuscles with inner core. A typical feature is that the Schwann cells form lamellae arranged to form an inner core. The corpuscle is enclosed in a perineural capsule. These nerve endings differ in size according to their position in the dermis (Poláček, 1968). The lower their position, the larger they are.

We believe that, with the aid of the electron microscope, the complex old nomenclature of nerve endings can be reduced to this simple system which is based exclusively on morphological considerations. Further research must determine whether this morphological system will correspond to a system based on functional studies.

D. The Mechanoreceptor Complexes of the Skin

The size, structure and number of the mechanoreceptors in the skin depends on the structure of the latter. The size of nerve endings depends on their depth below the epidermis (Poláček, 1968), their structure on the upper and lower surfaces of the epidermis (Halata, 1972c). The lower the nerve endings lie in the dermis, the larger they are, and the same type has different variants in hairy and hairless skin. We distinguished two types of skin in the glabrous skin of the mammals studied by us: cone skin (nose of mole and cat) and ridged skin (palmar digital skin of the rhesus monkey).

The nasal epidermis of the mole is arranged in approximately 5000 epidermal papillae (Eimer, 1871), consisting of a papillary elevation on the epidermal surface and an epidermal cone which is inverted into the dermis and has a cushion-like expansion. The nose of the cat shows a similar arrangement, except that the papillae on the epidermal surface are higher and the cones are thicker and penetrate more deeply into the dermis. The upper surface of the dermis (stratum papillare) mirrors the lower surface of the basal cell layer of the epidermis. The stratum papillare contains crypts, the width and depth of which correspond to the size and shape of the epidermal cones. The indentations are separated by connective tissue septa, which are penetrated by processes from the blood sinus lying deep in the dermis (nose of the mole). In the nose of the cat, the connective tissue septa contain capillaries which combine into a dense vascular bed.

There is a similarity between the innervation of the cone skin in the nose of the mole and that of the cat. The epidermal cones of the nose of the mole are innervated by free nerve endings and Merkel endings (Quilliam, 1966; Halata, 1972a). Each cone contains 20–22 free nerve endings and the base of the cone 3–5 Merkel endings. In the nose of the cat, the cones rarely contain free nerve endings. The base of the cone contains 5–15 Merkel endings. The stratum papillare under the cones contains encapsulated corpuscles with inner core (nose of the mole 1–3, nose of the cat 3–5). Simple bulboid nerve endings occur under the epidermal cones in the nose of the cat (Cauna's "papillary nerve endings", 1969). The structural elements of the cone skin, together with the nerve endings, may be regarded as forming a mechanoreceptor complex. A mechanoreceptor complex in the cone skin accordingly consists of the papillae rising above the epidermal surface, together with the stratum papillare of the dermis with its dense vascular bed, and the epidermal and dermal nerve endings.

The palmar skin of the monkey finger is arranged in ridges. As in man (Horstmann, 1957), the lower surface of the simian epidermis shows three types of ridges: glandular ridges, adhesive ridges, and cross ridges. The deepest and widest are the glandular ridges, which rise to form skin ridges on the epidermal surface. The surface ridges are separated by grooves corresponding to the adhesive ridges below. The cross ridges run perpendicular to the glandular and the adhesive ridges. This results in crypts on the lower surface of the epidermis, into which papillae extend from the stratum papillare of the dermis. The glandular ridges lie in vascular bed formed by capillaries of the stratum papillare of the dermis. The nerve endings of the ridged skin (Merkel endings in the bases of the glandular ridges and Meissner endings in the papillae of the dermis) form mechanoreceptor complexes together with the structural elements of the ridged skin (ridges and stratum papillare of the dermis with its vascular bed).

A typical example of a mechanoreceptor complex in hairy skin is represented by the vibrissa, or tactile or sinus hair. It differs from other hairs by its size and by the blood sinus (van Horn, 1970). The blood sinus surrounds the shaft and the bulb of the hair. The sinus hair has three types of mechanoreceptor: Merkel endings, bulboid nerve endings, and encapsulated corpuscles with inner core. 600–1000 Merkel endings are arranged in the form of a cuff in the basal layer of the sinus hair. The dendritic bulboid nerve endings (Andres's "lanciform endings", 1966a) are situated in the mesenchymal sheath of the sinus hair follicle. Encapsulated corpuscles with inner core (4–6) are situated deep in the mesenchymal sheath of the hair follicle. The sinus hair together with the mesenchymal sheath, the blood sinus and the capsule wall of the blood sinus, along with the above nerve endings, forms a highly sensitive touch apparatus. The receptors react to bending and twisting of the sinus hair (Gottschaldt, Iggo and Young, 1972). In all probability, the blood sinus insulates the nerve endings of the sinus hair against stimuli from the surrounding dermis (Andres, 1966a).

In the mechanoreceptor complexes of the hairless mammalian skin, the cones of the cone skin and the glandular ridges of the ridges skin may have a similar role to that of the sinus hair in transmitting mechanical stimuli. In the same way, the vascular bed, in which the cones and the glandular ridges are embedded, may have a similar function to that of the blood sinus of the sinus hair. Further studies will have to decide whether an impaired flow through these vascular beds would affect the response of the nerve endings in the mechanoreceptor complexes.

Summary

Light and electron microscopes were used to study the innervation of the glabrous skin of the nose of the mole (*Talpa europaea*), the cat (*Felis domestica*), the snout of the dwarf pig (Göttinger Miniaturschwein) the fingers of a monkey (*Maccacus rhesus*). We also described the innervation of the hair and sinus hair of the upper labial region of the cat.

We examined the mechanoreceptors in the mammalian epidermis and dermis and its related structures, and divided them into three types: Type I (epidermal nerve endings), Type II (bulboid nerve endings), and Type III (encapsulated corpuscles with inner core).

Type I includes the free nerve endings and the Merkel endings. Their distinguishing feature is the absence of a Schwann cell. The nerve terminal is expanded and lies between the epidermal cells. The afferent axons of the *free nerve endings* ($\emptyset$ 1–1.5 μ) are not myelinated in the dermis and are invested only with a Schwann cell and its basal lamina. On entering the epidermis, the axon loses its Schwann cell. The basal lamina of the Schwann cell fuses with that of the epidermis. In the epidermis, the axons are invaginated in the cells of the germinative zone but their course is extracellular. The terminal expansion lies in the granular layer. These endings were observed only in the epidermis of the glabrous nasal skin of the mole and the cat.

The Merkel endings consist of a terminal nerve disc and a Merkel cell. They lie between the basal layer and the stratum spinosum layer of the epidermal cone of the mole, the cat, and dwarf pig, and in the glandular ridge of the digital skin of the monkey. The afferent axon is myelinated in the dermis ($\emptyset = 2$–4 μ). After losing its myelin sheath, the axon, invested with a Schwann cell, approaches the basal lamina of the epidermis, it loses its Schwann cell. One myelinated axon can supply 2–8 Merkel cells. The Merkel cell differs from the keratinocytes by possessing a much-lobed nucleus and light cytoplasm, and by osmiophilic granules, which accumulate in the region of the cytoplasm, facing the nerve disc. There are synapse-like contacts between the nerve discs and the cytoplasmic membrane of the Merkel cell. The Merkel cell extends digitate processes between the keratinocytes. The Merkel cell is linked with the keratinocytes by desmosomes. The number of Merkel endings depends on the width and depth of the epidermal cones. The wider and deeper they are, the more Merkel endings they have.

Type II occurs in the dermis near the epidermis. It includes the simple and the dendritic bulboid nerve endings. The afferent axon of the simple bulboid nerve ending is not myelinated ($\emptyset = 2$ μ). Its terminals have bulboid expansions and are coated with a lamella of the Schwann cell and its basal lamina. The cell membrane of the Schwann cell is studded with pinocytotic vesicles. Unlike the afferent axon, the bulboid terminal contains accumulations of mitochondria and vesicles. The dendritic bulboid nerve endings include the palisade-form nerve endings of the hairs, the lanciform endings of the sinus hairs and the Meissner endings in the ridged skin. Unlike those of the simple bulboid nerve endings, the afferent axons are myelinated ($\varnothing = 3\text{--}5\,\mu$). The palisade-form nerve endings of the hair and the lanciform endings of the sinus hair run parallel to the longitudinal axis. The palisade-form nerve endings are in contact with the basal lamina of the hair. In the lanciform nerve endings of the sinus hair, the digitate processes of the nerve terminals in the upper region of the hair follicle are in contact with the basal lamina. In the lower part of the hair follicle, where the glassy membrane is thicker, the processes do not penetrate beyond the glassy membrane.

In the ridged skin, the terminals of the axons spiral towards the skin surface and so form Meissner endings in a dermal papilla. The terminals of the Meissner endings are enveloped by Schwann cell lamellae and their basal laminae. The nerve terminals and their Schwann cells are interspersed with collagen fibres and fibrocytes. The dendritic bulboid nerve endings have no perineural capsule.

Type III represents the encapsulated corpuscles with inner core. They consist of a nerve terminal, and inner core, a subcapsular space and a capsule. The afferent axon is myelinated ($\varnothing = 4\text{--}12\,\mu$). The nerve terminal is invested with a multi-layered lamellar system of Schwann cells and their basal laminae. This inner core contains two symmetrical longitudinal clefts. The encapsulated corpuscles of the mole form an exception, in that their inner core has no longitudinal clefts. The nerve terminal extends digitate processes into the spaces between the lamellae. The inner core is covered by a basal lamina. The inner core and the corpuscular capsule are separated by the subcapsular space, which contains collagen fibres and fibrocytes. The corpuscular capsule is formed by alternate layers of perineurial cells and collagen fibres. The individual layers of the perineurial cells are covered on both sides by a basal lamina. The number of layers depends on the size of the corpuscle. The capsule of the largest (Vater-Pacini) corpuscles contains up to 30 layers, those of the small corpuscles in the glabrous skin of the mole, one to two layers.

We examined the lower surface of the epidermis and its counterpart, the stratum papillare of the dermis. The glabrous skin of the mammals studied was divided into two types: cone skin and ridged skin.

The cone skin (mole, cat, dwarf pig): The base of the epidermis forms cones extending into the dermis. The cones correspond to the surface papillae of the horny layer of the epidermis. The cones are invaginated into indentations of the stratum papillare of the dermis. The individual cones are separated by the capillary bed of the dermis.

The ridged skin (monkey): The lower surface is formed by three types of ridges. The widest and deepest are the glandular ridges, which contain ducts of the sweat glands. The adhesive ridges run parallel to the glandular ridges, and cross ridges are perpendicular to both. This forms crypts on the lower surface of the epidermis which accomodate the papillae of the stratum papillare of the dermis. The glandular ridges have their surface counterpart in the skin ridges, and the adhesive ridges in the grooves.

The epidermis, the dermis and their mechanoreceptors form mechanoreceptor complexes. A typical example is represented by the sinus hair which contains all three types of nerve endings. The Merkel endings form a cuff at the level of the hair papilla. At the same level, the connective tissue of the inner hair follicle contains dendritic bulboid nerve endings (lanciform nerve endings). The lower third of the inner hair follicle contains encapsulated corpuscles with inner core. A blood sinus insulates the nerve endings against the surrounding dermis.

The mechanoreceptor complexes of the cone skin consists of an epidermal cone, the stratum papillare of the dermis, and their mechanoreceptors. The epidermal cones are separated by a capillary bed. In the ridged skin, the mechanoreceptor complexes are formed by the glandular ridges, the papillae of the dermis, and their mechanoreceptors. The glandular ridges are separated by a capillary network.

References

Andersen, A. E., Nafstad, P. H. J.: An electron-microscopic investigation of the sensory organs in the hard palate region of the hen (*Gallus domesticus*). Z. Zellforsch. **91**, 391–401 (1968)

Andres, K. H.: Über die Feinstruktur der Rezeptoren an Sinushaaren. Z. Zellforsch. **75**, 339–365 (1966a)

Andres, K. H.: Die Feinstruktur der sensiblen Endapparate gerader, zirculär und verzweigter Terminalfasern am Sinushaar. Naturwissenschaften 8, 204 (1966b)

Andres, K. H.: Zur Ultrastruktur verschiedener Mechanorezeptoren von höheren Wirbeltieren. Anat. Anz. **124**, 551–565 (1969)

Bielschowsky, M.: Über sensible Nervenendigungen in der Haut zweier Insectivoren. Anat. Anz. **31**, 187–196 (1907)

Böck, P.: Demonstration intraepithelialer Axone in der Papilla filiformis des Meerschweinchens. Acta anat. (Basel) **79**, 225–238 (1971)

Boeke, J.: Die interzelluläre Lage der Nervenendigungen im Epithelgewebe und ihre Beziehung zum Zellkern. Z. mikr.-anat. Forsch. **2**, 391–428 (1925)

Boeke, J.: Die Beziehungen der Nervenfasern zu den Bindegewebselementen und Tastzellen. Z. mikr.-anat. Forsch. **4**, 448–509 (1926)

Boeke, J.: Nerve endings, motor and sensory. In: Cytology and Cellular Pathology of the Nervous System. W. Penfield, pp. 243–315. New York: Hoerber 1932

Boeke, J.: Innervationsstudien. I. Noch einmal die intrazelluläre Lage der Nervenfasern und der Endknöpfchen im Epithel. Z. mikr.-anat. Forsch. **33**, 30–46 (1933a)

Boeke, J.: Innervationsstudien. II. Über Bau und Entwicklung des Eimerschen Organs in der Schnauze des Maulwurfs (*Talpa europaea*). Z. mikr.-anat. Forsch. **33**, 47–90 (1933b)

Boeke, J.: II. Niedere Sinnesorgane. 1. Freie Nervenendigungen und Endorgane sensibler Nerven. In: Handbuch der vergleichenden Anatomie der Wirbeltiere. Band 2, 2. Hälfte, S. 855–878. Berlin: Urban & Schwarzenberg 1934

Boeke, J.: Problems of Nervous Anatomy. Oxford: University Press 1940

Bonnet, J.: Studien über die Innervation der Haarbälge der Haustiere. Morph. Jb. **4**, 329–395 (1878)

Botezat, E.: Über das Verhalten der Nerven im Epithel der Säugetierenzunge. Z. wiss. Zool. **71**, 211–220 (1902a)

Botezat, E.: Die Nervenendigungen in der Schnauze des Hundes. Gegenbaurs morph. Jb. **29**, 439–449 (1902b)

Botezat, E.: Über die epidermoidalen Tastapparate in der Schnauze des Maulwurfs und anderer Säugetiere mit besonderer Berücksichtigung derselben für die Phylogenie der Haare. Arch. mikr. Anat. Entw.-Gesch. **61**, 730–764 (1903)

Botezat, E.: Die fibrilläre Struktur von Nervenendapparaten in Hautgebilden. Anat. Anz. **30**, 321–344 (1907)

Botezat, E.: Die Nerven der Epidermis. Anat. Anz. **33**, 45–75 (1908)

Botezat, E.: Die Apparate des Gefühlsinnes der nackten und behaarten Säugetierhaut, mit Berücksichtigung des Menschen. Anat. Anz. **42**, 193–205 und 273–318 (1912)

Breathnach, A. S.: An Atlas of the Ultrastructure of Human Skin. London: I. & A. Churchill 1971a

Breathnach, A. S.: Embryology of human skin. A review of ultrastructural studies. J. invest. Derm. **57**, 133–143 (1971b)

Breathnach, A. S., Robins, J.: Ultrastructural observation on Merkel cells in human fetal skin. J. Anat. (London) **106**, 411 (1970)

Burkel, W. E.: Perineurium, endoneurium and tissue space in peripheral nerve. Anat. Rec. **154**, 325–349 (1966)

Burkel, W. E.: The histological fine structure of perineurium. Anat. Rec. **158**, 177–190 (1967)

Cauna, N.: Some observations on the structure and development of Meissner's corpuscles. J. Anat. (London) **87**, 440–441 (1953)

Cauna, N.: Nature and functions of the papillary ridges of the digital skin. Anat. Rec. **119**, 449–468 (1954)

Cauna, N.: Structure and development of the capsule of Meissner's corpuscle. Anat. Rec. **124**, 77–94 (1956a)

Cauna, N.: Nerve supply and nerve endings of Meissner's corpuscles. Amer. J. Anat. **99**, 315–350 (1956b)

Cauna, N.: Structure of the digital touch corpuscles. Acta anat. (Basel) **32**, 1–23 (1968)

Cauna, N.: The mode of termination of the sensory nerves and its significance. J. comp. Neurol. **113**, 169–209 (1959)

Cauna, N.: Functional significance of the submicroscopical, histochemical and microscopical organisation of the cutaneous receptor organs. Erg.H. Anat. Anz. **111**, 181–197 (1962)

Cauna, N.: Fine structure of the receptor organs and its probable functional significance. In: Touch, heat and pain. Ciba Found. Symp., S. 117–127. London: I.& A. Churchill Ltd 1966

Cauna, N.: The fine morphology of the sensory receptor organs in the auricula of the rat. J. comp. Neurol. **136**, 81–98 (1969)

Cauna, N., Alberti, P.: Nerve supply and distribution of cholinesterase activity in the external nose of the mole. Z. Zellforsch. **54**, 158–166 (1961)

Cauna, N., Mannan, G.: The structure of human digital Pacinian corpuscles (corpuscula lamellosa) and its functional significance. J. Anat. (London) **92**, 1–20 (1958)

Cauna, N., Mannan, G.: Development and postnatal changes of digital Pacinian corpuscles in the human hand. J. Anat. (London) **93**, 271–286 (1959)

Cauna, N., Ross, L. L.: The fine structure of Meissner's touch corpuscles of human fingers. J. biophys. biochem. Cytol. 8, 467–482 (1960)

Chouchkov, N. H.: Ultrastructure of Pacinian corpuscles in men and cats. Z. mikr.-anat. Forsch. **83**, 17–32 (1971a).

Chouchkov, N. H.: Ultrastructure of Pacinian corpuscles after the section of nerve fibres. Z. mikr.-anat. Forsch. **83**, 33–46 (1971b)

Clara, M., Özer, N.: Untersuchungen über die sogenannte Nervenscheide. Acta neuroveg. (Wien) **20**, 1–18 (1959)

Cravioto, H.: The perineurium as a diffusion barrier. Ultrastructural correlates. Bull. Los Angeles neurol. Soc. **31**, 196–208 (1966)

Crevatin, F.: Über das strudelartige Geflecht der Hornhaut der Säugetiere. Anat. Anz. **19**, 411–413 (1901)

Cybulsky, I.: Das Nervensystem der Schnauze und der Oberlippe des Ochsen. Z. wiss. Zool. **39**, 18–32 (1883)

Dogiel, A. S.: Die Nervenendigungen in Tastkörperchen. Arch. Anat. Physiol., anat. Abt. 182–192 (1891)

Dogiel, A. S.: Die Nervenendigungen in Meissnerschen Tastkörperchen. Int. Mschr. Anat. Physiol. 9, 76–85 (1892)

Dogiel, A. S.: Die Nervenendigungen in der Haut der äußeren Genitalorgane des Menschen. Z. mikr.-anat. Forsch. **41**, 585–612 (1893)

Dogiel, A. S.: Über die Nervenapparate in der Haut des Menschen. Z. wiss. Zool. **75**, 46–110 (1903)

Dogiel, A. S.: Der fibrilläre Bau der Nervenendapparate in der Haut des Menschen und der Säugetiere und die Neuronentheorie. Anat. Anz. **27**, 97–120 (1905)

Duclaux, R., Kenshalo, D. R.: The temperature sensitivity of the type I slowly adapting mechanoreceptors in cats and monkeys. J. Physiol. (London) **224**, 647–664 (1972)

Eimer, T.: Die Schnauze des Maulwurfs als Tastwerkzeug. Arch. mikr. Anat. 7, 181–201 (1871)

Fischer, E.: Über den Bau der Meissnerschen Tastkörperchen. Arch. mikr. Anat. **12**, 364–390 (1876)

Fitzgerald, M. T. J.: Developmental changes in epidermal innervation. J. Anat. (London) **95**, 495–514 (1961)

Gamble, H. J.: Comparative electron microscopic observations on the connective tissues of a peripheral nerve and spinal nerve root in the rat. J. Anat. (London) **98**, 17–27 (1964)

Gamble, H. J.: Further electron microscopic studies of human peripheral nerves. J. Anat. (London) **100**, 487–502 (1966)

Gamble, H. J., Eames, R. A.: An electron microscopic study of the human peripheral nerves. J. Anat. (London) **98**, 655–663 (1964)

Gegenbauer, C.: Untersuchungen über die Tasthaare einiger Säugetiere. Z. wiss. Zool. **3**, 13–28 (1851)

Goethe, J.: Die intraepithelialen Nervenfasern im Vestibulum nasi der Kleinfedermäuse, dargestellt an Hand der Osmium-Zink-Jodid-Methode nach Maillet. Z. mikr.-anat. Forsch. **72**, 383–402 (1965)

Gottschaldt, K.-M., Iggo, A., Young, D. W.: Electrophysiology of the afferent innervation of sinus hairs, including vibrissae, of the cat. J. Physiol. (London) **222**, 60–61 P (1972)

Gussen, R.: Pacinian corpuscle in the middle ear. J. Laryng. **84**, 71–76 (1970)

Hagen, E.: Zur Innervation der Haut. Handbuch der Haut- und Geschlechtskrankheiten, Bd. I/1 (Erg.-Werk), S. 377–429. Berlin-Heidelberg-New York: Springer 1968

Hagen, E., Werner, S.: Beobachtungen an der Ultrastructur epidermaler und subepidermaler Nerven vor und nach Röntgenbestrahlung. I. Mitt. Arch. klin. exp. Derm. **225**, 306–327 (1966a)

Hagen, E., Werner, S.: Beobachtungen an der Ultrastruktur epidermaler und subepidermaler Nerven vor und nach Röntgenbestrahlung. II. Mitt. Arch. klin. exp. Derm. **225**, 328–334 (1966b)

Hagen, E., Werner, S.: Zur Ultrastruktur des Nervensystems in der Haut. Erg.H. z. Anat. Anz. **120**, 277–288 (1967)

Halata, Z.: Zu den Nervenendigungen (Merkelsche Endigungen) in der haarlosen Nasenhaut der Katze. Z. Zellforsch. **106**, 51–60 (1970)

Halata, Z.: Ultrastructure of Grandry nerve endings in the beak skin of some aquatic birds. Folia Morph. (Prague) **19**, 225–232 (1971a)

Halata, Z.: Die Ultrastruktur der Lamellenkörperchen bei Wasservögeln (Herbstsche Endigungen). Acta anat. (Basel) **80**, 362–376 (1971b)

Halata, Z.: Innervation der unbehaarten Nasenhaut des Maulwurfs (*Talpa europaea*). I. Intraepidermale Nervenendigungen. Z. Zellforsch. **125**, 108–120 (1972a)

Halata, Z.: Innervation der unbehaarten Nasenhaut des Maulwurfs (*Talpa europaea*). II. Innervation der Dermis (einfache eingekapselte Körperchen). Z. Zellforsch. **120**, 121–131 (1972b)

Halata, Z.: Über Zusammenhänge zwischen der Hautoberfläche und dem Tastapparat der Haut bei verschiedenen Säugern. Verh. anat. Ges. (Jena) **67**, 459–466 (1973)

Hashimoto, K.: Merkel tactile cells of the human embryo nail. 29th Ann. Proc. of El. Micr. Soc. of America **1971**, 546–547

Hashimoto, K.: The ultrastructure of the skin of human embryos. X. Merkel tactile cells in the finger and nail. J. Anat. (London) **111**, 99–120 (1972a)

Hashimoto, K.: Fine structure of Merkel cell in human oral mucosa. J. invest. Derm. **58**, 381–387 (1972b)

Hashimoto, K.: Fine structure of perifollicular nerve endings in human hair. J. invest. Derm. **59**, 432–441 (1972c)

Hebel, R., Schweiger, A.: Zur Feinstruktur und Funktion sensibler Rezeptoren. Zbl. Vet.-Med. A **14**, 15–24 (1967)

Henle, J., Kölliker, A.: Über die Pacinischen Körperchen an den Nerven des Menschen und der Säugetiere. Shortened English translation in the British and Foreign Medical Review **19**, 78–83. London: Churchill 1845

Hensel, H.: Cutane Wärmerezeptoren bei Primaten. Pflügers Arch. Eur. J. Physiol. **313**, 150–152 (1969)

Hensel, H., Iggo, A.: Analysis of cutaneous warm and cold fibers in primates. Pflügers Arch. Eur. J. Physiol. **329**, 1–8 (1971)

Heringa, G. A.: Untersuchungen über den Bau und die Entwicklung des sensiblen peripheren Nervensystems. Verh. kon. Akad. Wet. Amsterdam **21**, 1–30 (1920)

Hoggan, G.: New forms of nerve terminations in mammalian skin. J. Anat. Physiol. **18**, 182–197 (1884)

Horstmann, E.: Die Haut. In: v. Möllendorff, Handbuch der mikroskopischen Anatomie des Menschen; Haut und Sinnesorgane III/3, S. 1–276. Berlin-Göttingen-Heidelberg: Springer 1957

Huss, G.: Beiträge zur Kenntnis der Eimerschen Organe in der Schnauze von Säugern. Z. wiss. Zool. **67**, 102–127 (1900)

Huxley, T. H.: On the structure and relation of the Corpuscula Tactus (tactile corpuscles or axiale corpuscles) and of the Pacini bodies. Quart. J. micr. Sci. **2**, 1–7 (1854)

Iggo, A.: New specific sensory structures in hairy skin. Acta neuroveg. (Wien) **24**, 175–180 (1963)

Iggo, A.: Cutaneous receptors with a high sensitivity to mechanical displacement. In: Touch, heat and pain; Ciba Symp. Found., S. 237–256. London: A. Churchill 1966

Iggo, A.: Cutaneous thermoreceptors in primates and subprimates. J. Physiol. (London) **200**, 403–430 (1969)

Iggo, A., Muir, A. R.: A cutaneous sense organ in the hairy skin of cats. J. Anat. (London) **97**, 151 (1963)

Iggo, A., Muir, A. R.: The structure and function of slowly adapting touch corpuscles in hairy skin. J. Physiol. (London) **200**, 763–796 (1969)

Ito, S., Winchester, R. J.: The fine structure of the gastric mucosa in the bat. J. Cell Biol. **16**, 541–578 (1963)

Ivanov, G. F.: The morphology and classification of some forms of interoreceptors (Russian text). In: Problemi kortikovisceral'noi patologii. Moscow, 135–153 (1949)

Jabonero, V., Moya, J.: Studien über die sensiblen Ausbreitungen. Acta anat. (Basel) **78**, 488–520 (1971)

Jaburek, A. L.: Die Nervenendigungen in der Epidermis der Reptilien. Z. mikr.-anat. Forsch. **10**, 1–49 (1927)

Jalowy, B.: Über die Degeneration und Regeneration der Nervenendigungen in den Fingerbeeren der oberen Extremität der Affen (*Macacus rhesus*). Z. Zellforsch. **23**, 84–116 (1936)

Jänig, W.: Morphology of rapidly and slowly adapting mechanoreceptors in the hairless skin of the cats hind foot. Brain Res. **28**, 217–231 (1971)

Kadanoff, D.: Eine besondere Nervenendigung in der Haut des Menschen. Z. Anat. Entwickl.-Gesch. **72**, 542–544 (1924a)

Kadanoff, D.: Beitrag zur Kenntnis der Nervenendigungen im Epithel der Säugetiere. Z. Anat. Entwickl.-Gesch. **73**, 431–452 (1924b)

Kadanoff, D.: Über die intraepithelialen Nerven und ihre Endigungen beim Menschen und bei den Säugetieren. Z. Zellforsch. **7**, 553–576 (1928)

Kadanoff, D.: Über intraepitheliale nervöse Endformationen in mehrschichtigem Plattenepithel. Z. Zellforsch. **121**, 171–180 (1971a)

Kadanoff, D.: Die Lagebeziehungen der intraepithelialen Nervenfasern und -Endigungen zu den Epithelzellen. Compt. rend. Ac. Bulg. Sci. **24**, 941–944 (1971b)

Kadanoff, D.: Die Ultrastruktur und Lage der Nervenfasern und ihrer Endigungen im Epithelgewebe. Z. mikr.-anat. Forsch. **84**, 321–332 (1971c)

Kadanoff, D., Seguchi, H., Villiger, W.: Ultrastructural investigations of the palisadeshaped nerve fiber terminals of the normal hairs of rat's snout. Z. Zellforsch. **147**, 259–269 (1974)

Kasprzak, H., Tapper, D. N., Craig, P. H.: Functional development of the tactile pad receptor system. Exp. Neurol. **26**, 439–446 (1970)

Kawamura, T.: Über die menschliche Haarscheibe, unter besonderer Berücksichtigung ihrer Innervation und subepidermalen perineuralen Pigmenthülle. Hautarzt **5**, 106–109 (1955)

Kawamura, T., Nishiyma, S., Ikeda, S.: The human Haarscheibe, its structure and function. J. invest. Derm. **42**, 87–90 (1964)

Kerjaschki, D., Stockinger, L.: Structure and function of the perineurium. Z. Zellforsch. **110**, 386–400 (1970)

Key, A., Retzius, G.: Studien in der Anatomie des Nervensystems und des Bindegewebes. Stockholm: Samson and Wallin 1876

Kidd, R. L., Krawczyk, W. S., Wilgram, G. F.: The Merkel cell in human epidermis: its differentiation from other dendritic cells. Arch. Derm. Forsch. **241**, 374–384 (1971)

Kölliker, A.: Über den Bau der Cutispapillen und der sogenannten Tastkörperchen R. Wagners. Z. wiss. Zool. **4**, 43–52 (1852)

Krause, W.: Die Terminalkörperchen der einfachen sensiblen Nerven. Hannover: Hansche Hofbuchhandlung 1860

Krause, W.: Die Nervenendigungen innerhalb der terminalen Körperchen. Arch. mikr. Anat. 19, 53–136 (1881)

Ksjunin, W.: Zur Frage über die Nervenendigungen in den Tast- und Sinushaaren. Arch. mikr. Anat. 54, 403–420 (1899)

Langerhans, P.: Über die Nerven der menschlichen Hand. Virchows Arch. path. Anat. 44, 325–337 (1868)

Lavrentjev, B. I.: Sensory innervation of inner organs (Russian text). In: Morfol. cuvst. innerv. vnutr. organov. 1947, 5–21

Lefébure, M.: Les corpuscles de Wagner-Meissner ou corpuscles du tact. Rev. gen. Histol. 3, 569–736 (1910)

Lehmann, H. J.: Über Struktur und Function der perineuralen Diffusionsbarriere. Z. Zellforsch. 46, 223–241 (1957)

Lele, P. P., Weddel, G.: Sensory nerves of the cornea and cutaneous sensibility. Exp. Neurol. 1, 334–360 (1959)

Leontowitsch, A.: Die Innervation der menschlichen Haut. Int. Mschr. Anat. Physiol. 18, 142–310 (1901)

Levi, S.: Osservazioni sullo svilupo delle terminazioni nervose intraepithelali, corpuscoli de Meissner a corpuscoli del Pacini. Arch. Italiano di Anat. e di Embriol. 32, 149–170 (1933

Lipetz, L. E.: The relation of physiological and psychological aspects of sensory intensity In: Principles of Receptor Physiology, p. 191–225. Ed. W. R. Loewenstein. Berlin-Heidelberg-New York: Springer 1971

Loewenstein, W. R.: Mechano-electric transduction in the Pacinian corpuscle. Initiation of sensory impulses in mechanoreceptors. In: Principles of Receptor Physiology, p. 269–290. Ed. W. R. Loewenstein. Berlin-Heidelberg-New York: Springer 1971

Loo, S. K., Kanagasuntheram, R.: Innervation and structure of the snout in the three shrew. J. Anat. (London) 111, 253–261 (1972)

Luft, J. H.: Improvements in epoxy resin embedding methods. J. biophys. biochem. Cytol. 9, 409–414 (1961)

Lyne, A. G., Hollis, D. E.: Merkel cells in the sheep epidermis during fetal development. J. Ultrastruct. Res. 34, 464–472 (1971)

Malinovský, L.: Variability of sensory nerve endings in foot pads of a domestic cat (Felis ocreata L., F. domestica). Acta anat. (Basel) 64, 82–106 (1966a)

Malinovský, L.: The variability of encapsulated corpuscles in the upper lip and tongue of the domestic cat. Folia Morph. (Prague) 14, 175–191 (1966b)

Malinovský, L.: Variability of sensory corpuscles in the skin of the nose and in the area of sulcus labii maxillaris of the domestic cat. Folia Morph. (Prague) 14, 417–429 (1966c)

Malinovský, L.: Ein Beitrag zur Entwicklung einfacher sensibler Körperchen bei der Hauskatze (Felis silvestris, f. catus L.). Acta anat. (Basel) 76, 220–235 (1970)

Mann, S. J., Straile, W. E.: Tylotrich (hair) follicle: Association with a slowly adapting receptor in the cat. Science 147, 1043–1045 (1965)

McGavran, M. H.: "Chromaffin" cell: electron microscopic identification in the human dermis. Science 145, 275–276 (1964)

Meissner, G.: Beiträge zur Anatomie und Physiologie der Haut. Leipzig: Leopold Voss 1853

Meissner, G.: Bemerkungen die Tastkörperchen betreffend. Z. wiss. Zool. 6, 296–297 (1855)

Meissner, G.: Untersuchungen über den Tastsinn. Z. rat. Med. Reihe 7, 92–119 (1895)

Melaragno, H. P., Montagna, W.: The tactile hair follicles in the mouse. Anat. Rec. 115, 129–150 (1953)

Merkel, F.: Tastzellen und Tastkörperchen bei den Haustieren und beim Menschen. Arch. mikr. Anat. 11, 636–652 (1875)

Merkel, F.: Die Tastzellen der Ente. Arch. mikr. Anat. 15, 415–427 (1878)

Merkel, F.: Über die Endigungen der sensiblen Nerven in der Haut der Wirbeltiere. Rostock: Schmidt 1880

Miller, M. R., Ralston III, H. J., Kasahara, M.: The pattern of cutaneous innervation of the human hand. Amer. J. Anat. 102, 183–218 (1958)

Miller, M. R., Ralston III, H. J., Kasahara, M.: The pattern of cutaneous innervation of the human hand, foot and breast. In: Advances in Biology of Skin, vol. I, Cutaneous Innervation, p. 1–47. Ed. W. Montagna. New York: Pergamon Press 1960

Mojsisovics: Über die Nervenendigungen in der Epidermis der Säuger. II. S.B. Akad. wiss. Wien, math-nat. Kl. 1876, 73

Montagna, W.: The structure and function of skin. New York and London: Academic Press 1962

Munger, B. R.: The intraepidermal innervation of the snout skin of the opossum. A light and electron microscope study, with observations on the nature of Merkel's Tastzelle. J. Cell Biol. **26**, 79–97 (1965)

Munger, B. L.: Diskussionsbeitrag. In: Touch, heat and pain. Ciba Found. Symp., p. 121–132. London: I. & A. Churchill Ltd. 1966a

Munger, B. L.: The ultrastructure of Herbst and Grandry corpuscles. Anat. Rec. **154**, 394–392 (1966b)

Munger, B. L.: Patterns of organisation of periphery sensory receptors. In: Principles of Receptor Physiology, p. 523–556. Ed. W. Loewenstein. Berlin-Heidelberg-New York: Springer 1971

Munger, B. L., Pubols, L. M., Pubols, B. H.: The Merkel rete papilla—a slowly adapting sensory receptor in mammalian glabrous skin. Brain Res. **29**, 47–61 (1971)

Mustakallio, K. K., Kiistala, U.: Electron microscopy of Merkel's "Tastzelle", a potential monoamine storing cell of human epidermis. Acta derm. venereol. (Stockh.) **47**, 323–326 (1967)

Nafstad, P. H. J.: Comparative ultrastructural study on Merkel cells and dermal basal cells in poultry (*Gallus domesticus*). Z. Zellforsch. **116**, 342–348 (1971a)

Nafstad, P. H. J.: On the ultrastructure of neuro-epithelial interactions. The dermal innervation in the snout of the pig. Z. Zellforsch. **122**, 528–537 (1971b)

Nafstad, P. H. J.: On the dermal innervation. Z. Anat. Entwickl.-Gesch. **135**, 337–349 (1972)

Niculescu, I.: Atlas privid aspecte morfologice ale terminatiilir nervoase viscerale. Editura medical, Bucuresti, 1–455 (1958)

Niewöhner, R. E., van der Zypen, E.: Über die sensible Innervation der Nasenregion der Katze. Anat. Anz. **130**, 70–90 (1972)

Nikai, H., Rose, G. G., Cattoni, M.: Merkel cell in human and rat gingiva. Arch. oral Biol. **16**, 835–844 (1971)

Nishi, K., Oura, C., Pallie, W.: Fine structure of Pacinian corpuscles in the mesentery of the cat. J. Cell Biol. **43**, 539–552 (1969)

Novotný, V., Halata, Z.: A study of the joint receptors of the bat as a flying mammal. Folia Zool. (Prague) (im Druck)

Ono, S.: Histologic study on the innervation of the snout and the nasal cavity with their surroundings in cat. Arch. histol. jap. **10**, 37–52 (1956)

Orfanos, C.: Elektronenmikroskopische Befunde an epidermisnahen Nervenanteilen. Arch. klin. exp. Derm. **222**, 603–612 (1965a)

Orfanos, C.: Der Aufbau der peripheren Nervenfaser der menschlichen Haut. Arch. klin. exp. Derm. **223**, 457–477 (1965b)

Orfanos, C.: Elektronenmikroskopischer Nachweis epithelioneuraler Verbindungen am Haarfollikelepithel des Menschen. Arch. klin. exp. Derm. **228**, 421–429 (1967)

Ormea, F., Goglia, G.: Ultrastructural research of the Krause's nerve endings (zylindrische Endkolben und kugelige Endkolben). Ital. gen. Rev. Derm. **9**, 1–22 (1969)

Ostroumow, M. P.: Die Nervenendigungen der Sinushaare. Anat. Anz. **10**, 781–790 (1895)

Pacini, F.: Sopra un particolare genere di piccoli Corpi globulosi scorpeti nel corpo umano da Filippo Pacini Aluno interna degli spedali riunti di Pistoia. (Letter to Accademia Medicofisica di Firenze 1835)

Pallie, W., Nishi, K., Oura, C.: The pacinian corpuscle, its vascular supply and the inner core. Acta anat. (Basel) **77**, 508–520 (1970)

Patrizi, G., Munger, B. L.: The structure and innervation of rat vibrissae. J. comp. Neurol. **126**, 423–436 (1966)

Pease, D. C., Pallie, W.: Electron microscopy of digital tactile corpuscle and small cutaneous nerves. J. Ultrastruct. Res. **2**, 352–362 (1959)

Pease, D. C., Quilliam, T. A.: Electron microscopy of the Pacinian corpuscle. J. biophys. biochem. Cytol. **3**, 331–342 (1957)

Pieper, A.: Die interzelluläre Lage der Nerven im Epithel. Anat. Anz. **91**, 288–312 (1941)

Pincus, F.: Über bisher unbekannten Nervenapparat am Haarsystem des Menschen: Haarscheiben. Derm. Z. **9**, 465–469 (1902)

Pincus, F.: Über Hautsinnesorgane neben dem menschlichen Haar (Haarscheiben) und ihre vergleichsanatomische Bedeutung. Arch. mikr. Anat. **65**, 121–179 (1905)

Podcopäev: Über die Endigung der Nerven in der epithelialen Schicht der Haut. Arch. mikr. Anat. **5**, 506–508 (1869)

Poláček, P.: Differences in the structure and variability of encapsulated nerve endings in the joint of some species of some mammals. Acta anat. (Basel) **47**, 112–124 (1961)

Poláček, P.: Differences in the structure and variability of spray-like nerve endings in the joint of some mammals. Acta anat. (Basel) **62**, 568–583 (1965)

Poláček, P.: Receptors of the joint, their structure, variability and classification. Acta Fac. Med. Univ. Brun **23**, 1–107 (1966)

Poláček, P.: Über die strukturellen Unterschiede der Rezeptorreihen in der Vaginalwand der Katze und ihre mögliche funktionelle Bedeutung. Z. mikr.-anat. Forsch. **78**, 1–34 (1968)

Poláček, P.: Die Ultrastruktur des Herbstschen Körperchen im Vergleich mit dem Vater-Pacinischen Körperchen (vorläufige Mitteilung). Coll. Sci. Works Fac. Med., Charles Univ. Hradec Králové **12**, 417–426 (1969)

Poláček, P.: Některé nové pohledy na morfologii sensitivních nervových zakončení. Lék. Zprávy (Hradec Králové) **16**, 9–13 (1971)

Poláček, P., Halata, Z.: Beziehung der Kapsel der Nervenendigungen zu den Hüllen des Nervensystems. Scr. med. Fac. Med. Brun. **38**, 73–83 (1965)

Poláček, P., Halata, Z.: Development of simple encapsulated corpuscles in the nasolabial region of the cat (Ultrastructural study). Folia Morphol. (Prague) **18**, 359–368 (1970)

Poláček, P., Malinovský, L.: Die Ultrastruktur der Genitalkörperchen in der Clitoris. Z. mikr.-anat. Forsch. **84**, 293–310 (1971)

Poláček, P., Mazanec, K.: Ultrastructure of mature Pacinian corpuscle from mesentery of adult cat. Z. mikr.-anat. Forsch. **74**, 343–354 (1966)

Quilliam, T. R.: Structure of receptor organs. Unit design and array patterns in receptor organs. In: Touch, heat and pain. Ciba Found. Symp., p. 86–116. London: I. & A. Churchill Ltd. 1966

Quilliam, T. R., Armstrong, J.: Mechanorezeptoren. Endeavour **12**, 55–60 (1963)

Rachmatulin, Z. Ch.: Die Entwicklung der Meissnerschen Körperchen in der Menschenhaut. Z. mikr.-anat. Forsch. **40**, 445–454 (1936)

Ranvier, L.: On the terminations of nerves in the epidermis. Quart. J. micr. Sci. **20**, 456–458 (1880a)

Ranvier, L.: Nouvelles recherches sur organes du tact. C.R. Acad. Sci. **41**, 1087–1089 (1880b)

Retzius, G.: Über die sensiblen Nervenendigungen in den Epithelien der Wirbeltiere. Biol. Unters. **4**, 37–44 (1892)

Retzius, G.: Einige Beiträge zur Kenntnis der intraepithelialen Endigungsweise der Nervenfasern. Biol. Unters. **6**, 62–64 (1894)

Reynolds, E. S.: The use of lead citrate at high pH as an electron-opaque stain in electron microscopy. J. Cell Biol. **17**, 208–212 (1963)

Ridley, A.: Silver staining of nerve endings in human glabrous skin. J. Anat. (London) **104**, 41–48 (1969).

Röhlich, P.: Submicroscopical morphology of the Perineurium. Acta Morphol. Hung., Suppl. **8**, 47–62 (1959)

Röhlich, P., Knoop, A.: Elektronenmikroskopische Untersuchungen an den Hüllen des Nervus ischiadicus der Ratte. Z. Zellforsch. **53**, 299–312 (1961)

Röhlich, P., Weiss, M.: Studies on the history and permeability of the peripheral nervous barrier. Acta Morphol. Hung. **5**, 335–347 (1955)

Ross, M. H., Reith, E. J.: Perineurium: Evidence for contractile elements. Science **165**, 604–605 (1969)

Sasybin, N. I.: Über die Regeneration der Nervenfasern in mehrschichtigem Plattenepithel. Z. mikr.-anat. Forsch. **22**, 1–72 (1930)

Schmidt, R. F.: Möglichkeiten und Grenzen der Hautsinne. Klin. Wschr. **49**, 530–540 (1971)

Shanthaveerappa, T. R., Bourne, G. H.: A perineural epithelium. J. biophys. biochem. Cytol. **14**, 343–346 (1962a)

Shanthaveerappa, T. R., Bourne, G. H.: The "perineural epithelium" a metabolically active, continuous, protoplasmic cell barrier surrounding peripheral nerve fasciculi. J. Anat. (London) **96**, 527–537 (1962b)

Shanthaveerappa, T. R., Bourne, G. H.: Demonstration of perineural epithelium in whale and shark peripheral nerves. Nature (London) **197**, 702–703 (1963a)

Shanthaveerappa, T. R., Bourne, G. H.: The perineural epithelium: Nature and significance. Nature (London) **199**, 577–579 (1963b)

Shanthaveerappa, T. R., Bourne, G. H.: Demonstration of perineural epithelium in vagus nerve. Acta anat. (Basel) **52**, 95–100 (1963c)

Shanthaveerappa, T. R., Bourne, G. H.: New observations on the structure of the Pacinian corpuscle and its relation to the perineural epithelium of peripheral nerves. Amer. J. Anat. **112**, 97–109 (1963d)

Shanthaveerappa, T. R., Bourne, G. H.: The perineural epithelium of sympathetic nerves and ganglia and its relation to the pia-arachnoid of the central nervous system and perineural epithelium of the peripheral nervous system. Z. Zellforsch. **61**, 742–753 (1964a)

Shanthaveerappa, T. R., Bourne, G. H.: The effects of transsection of the nerve trunk of the perineural epithelium with special reference to its role in nerve degeneration and regeneration. Anat. Rec. **150**, 35–50 (1964b)

Shanthaveerappa, T. R., Hope, J., Bourne, G. H.: Electron microscopic demonstration of perineural epithelium in rat peripheral nerve. Acta anat. (Basel) **52**, 193–201 (1963)

Shehata, R.: Pacinian corpuscles in bladder wall and outside ureter of the cat. Acta anat. (Basel) **77**, 139–143 (1970)

Shehata, R.: Pacinian corpuscles in pelvic urogenital organs and outside abdominal lymph glands of the cat. Acta anat. (Basel) **83**, 127–138 (1972)

Sinclair, D.: The nerve endings. In: Cutaneous sensation, p. 35–56. New York: Oxford University Press 1967

Smirnow, A.: Über Endkolben in der Haut der Planta pedis und über die Nervenendigungen in den Tastkörperchen des Menschen. Mnt.schr. Anat. Physiol. **10**, 241–249 (1893)

Smith, K. R.: The structure and function of Haarscheibe. J. comp. Neurol. **131**, 459–474 (1967)

Smith, K. R.: The ultrastructure of human "Haarscheibe" and Merkel cell. J. invest. Derm. **54**, 150–159 (1970)

Smith, K. R., Creech, B. J.: Effects of a pharmacological agents on the physiological responses of hair discs. Exp. Neurol. **19**, 477–482 (1965)

Spassova, I.: Ultrastructure of the encapsulated receptors in the lips of the cat. C.R. Ac. Bulgare Sci. **23**, 1311–1313 (1970)

Spassova, I.: Ultrastructure of Krause end-bulbs in the nasal skin of the cat. Acta anat. (Basel) **84**, 224–236 (1973)

Stöhr, P. , Jr.: Das peripherische Nervensystem. In: v. Möllendorffs Handbuch der mikroskopischen Anatomie des Menschen, Vol. 4, erster Teil, S. 143–264. Berlin: Springer 1928

Straile, W. E.: Sensory hair follicles in mammalian skin: The tylotrichic follicle. Amer. J. Anat. **106**, 133–148 (1960)

Szymonowicz, L.: Beiträge zur Kenntnis der Nervenendigungen in Hautgebilden. Arch. mikr. Anat. **45**, 624–653 (1895)

Szymonowicz, L.: Über den Bau und Entwicklung der Nervenendigungen im Entenschnabel. Arch. mikr. Anat. **48**, 329–358 (1897)

Szymonowicz, L.: Über die Entwicklung der Nervenendigungen in der Haut des Menschen. Z. Zellforsch. **19**, 356–382 (1933)

Thomas, P. K.: The connective tissue of peripheral nerve, an electron microscopic study. J. Anat. (London) **97**, 35–44 (1963)

Tomsa, W.: Zur Kenntnis der Nervenenden in der Haut der menschlichen Hand. Wien. med. Wschr. **53**, 793–796 (1865)

Tretjakoff, D.: Zur Frage der Nerven der Haut. Z. wiss. Zool. **71**, 625–644 (1902)

Tretjakoff, D.: Die Nervenendigungen in den Sinushaaren des Rindes. Z. wiss. Zool. **97**, 314–416 (1911)

Tsuji, T.: Free nerve endings of the epidermis in hairy and hairless mice. J. invest. Derm. **57**, 247–255 (1971)

Van der Velde, E.: Die fibrilläre Struktur der Nervenenden. Int. Mschr. Anat. Physiol. **24**, 225–298 (1909)

Van Horn, R. N.: Vibrissae structure in the rhesus monkey. Folia primat. **13**, 241–285 (1970)

Vater, A.: Dissertation de consensu partium corporis humani. Haller, Disputationum anatomicarum selectarum, vol. II, Gottingae 1741

Vincent, S. B.: The tactile hair of the white rat. J. comp. Neurol. **23**, 1–38 (1913)

Waggener, J. P., Beggs, J.: The membranous coverings of neural tissues: An electron microscopic study. J. Neuropath. exp. Neurol. **26**, 412–426 (1967)

Waggener, J. P., Bunn, M. S., Beggs, J.: The diffusion of ferritin within the peripheral nerve sheath. An electron microscopic study. J. Neuropath. exp. Neurol. **24**, 430–443 (1965)

Wagner, R., Meissner, G.: Über das Vorhandensein bisher unbekannter eigenthümlicher Tastkörperchen (corpuscula tactus) in den Gefühlswärzchen der menschlichen Haut und über
die Endausbreitung sensitiver Nerven. Nachrichten Georg-August-Uni. königl. Ges. Wiss.
Göttingen 2, 17–30 (1852)

Walter, P.: Zur Innervation der Pferdelippe. Z. Zellforsch. 43, 459–477 (1956)

Walter, P.: Die sensible Innervation des Lippen- und Nasenbereiches von Rind, Schaf, Ziege,
Schwein, Hund und Katze. Z. Zellforsch. 53, 394–410 (1960/61)

Walter, P.: Licht- und elektronenmikroskopische Untersuchungen an sensiblen Rezeptoren
von Haustieren. Erg.H. z. Anat. Anz. 111, 198–207 (1962)

Walter, P., Hebel, R.: Zur Morphologie sensibler Rezeptoren. Anat. Anz. 118, 436–443 (1966)

Weddel, G.: The anatomy of cutaneous sensibility. Brit. Med. Bull. 3, 167–172 (1945)

Weddel, G., Pallie, W., Palmer, E.: The morphology of peripheral nerve terminations in the
skin. Quart. J. micr. Sci. 95, 483–501 (1954)

Winkelmann, R. K.: The sensory end-organs of the hairless skin of the cat. J. invest. Derm.
29, 347–352 (1957a)

Winkelmann, R. K.: The muco-cutaneous end-organ. Arch. Derm. 76, 225–235 (1957b)

Winkelmann, R. K.: The sensory nerve endings in the skin of the cat. J. comp. Neurol. 109,
221–232 (1958)

Winkelmann, R. K.: Similarities in cutaneous nerve end-organs. In: Advances in biology of
skin, vol. I. Cutaneous Innervation, p. 48–62. Ed. W. Montagna. New York: Pergamon
Press 1960a

Winkelmann, R. K.: The end-organ of feline skin: A morphologic and histochemic study.
Amer. J. Anat. 107, 281–290 (1960b)

Woolard, H. H.: Intra-epidermal nerve endings. J. Anat. (London) 71, 54–60 (1933)

Zabusov, G. I., Maslov, A. P.: A proposition of evolutionary and morphological classification
of sensory nerve endings (Russian Text). In: Problemi morfologii, patomorfologii i reaktiv-
nosti perifericeskich otdelov nervnoj sistemi; Kazan, 41–58 (1961)

Zollmann, P. E., Winkelmann, R. K.: The sensory innervation of common North American
racoon (Procyon lotor). J. comp. Neurol. 119, 149–158 (1962)

Subject Index

Advances in Anatomy, Embryology and Cell Biology
Ergebnisse der Anatomie und Entwicklungsgeschichte
Revues d'anatomie et de morphologie expérimentale
Springer-Verlag   Berlin   Heidelberg   New York

This journal publishes reviews and critical articles covering the entire field of normal anatomy (cytology, histology, cyto- and histochemistry, electron microscopy, macroscopy, experimental morphology and embryology and comparative anatomy). Papers dealing with anthropology and clinical morphology will also be accepted with the aim of encouraging co-operation between anatomy and related disciplines.

Papers, which may be in English, French or German, are normally commissioned, but original papers and communications may be submitted and will be considered so long as they deal with a subject comprehensively and meet the requirements of the Ergebnisse.

For speed of publication and breadth of distribution, this journal appears in single issues which can be purchased separately; 6 issues constitute one volume.

It is a fundamental condition that submitted manuscripts have not been, and will not simultaneously be submitted or published elsewhere. With the acceptance of a manuscript for publication, the publishers acquire full and exclusive copyright for all languages and countries.

25 copies of each paper are supplied free of charge.

Les résultats publient des sommaires et des articles critiques concernant l'ensemble du domaine de l'anatomie normale (cytologie, histologie, cyto et histochimie, microscopie électronique, macroscopie, morphologie expérimentale, embryologie et anatomie comparée. Seront publiés en outre les articles traitant de l'anthropologie et de la morphologie clinique, en vue d'encourager la collaboration entre l'anatomie et les disciplines voisines.

Seront publiés en priorité les articles expressément demandés nous tiendrons toutefois compte des articles qui nous seront envoyés dans la mesure où ils traitent d'un sujet dans son ensemble et correspondent aux standards des «Résultats». Les publications seront faites en langues anglaise, allemande et française.

Dans l'intérêt d'une publication rapide et d'une large diffusion lestravaux publiés paraitront dans des cahiers individuels, diffusés séparément: 6 cahiers forment un volume.

En principe, seuls les manuscrits qui n'ont encore été publiés ni dans le pays d'origine ni à l'étranger peuvent nous être soumis. L'auteur d'engage en outre à ne pas les publier ailleurs ultérieurement.

Les auteurs recevront 25 exemplaires gratuits de leur publication.

Die Ergebnisse dienen der Veröffentlichung zusammenfassender und kritischer Artikel aus dem Gesamtgebiet der normalen Anatomie (Cytologie, Histologie Cyto- und Histochemie, Elektronenmikroskopie, Makroskopie, experimentelle Morphologie und Embryologie und vergleichende Anatomie). Aufgenommen werden ferner Arbeiten anthropologischen und morphologisch-klinischen Inhaltes, mit dem Ziel, die Zusammenarbeit zwischen Anatomie und Nachbardisziplinen zu fördern.

Zur Veröffentlichung gelangen in erster Linie angeforderte Manuskripte, jedoch werden auch eingesandte Arbeiten und Originalmitteilungen berücksichtigt, sofern sie ein Gebiet umfassend abhandeln und den Anforderungen der ,,Ergebnisse" genügen. Die Veröffentlichungen erfolgen in englischer, deutscher und französischer Sprache.

Die Arbeiten erscheinen im Interesse einer raschen Veröffentlichung und einer weiten Verbreitung als einzeln berechnete Hefte; je 6 Hefte bilden einen Band.

Grundsätzlich dürfen nur Arbeiten eingesandt werden, die nicht gleichzeitig an anderer Stelle zur Veröffentlichung eingereicht oder bereits veröffentlicht worden sind. Der Autor verpflichtet sich, seinen Beitrag auch nachträglich nicht an anderer Stelle zu publizieren.

Die Mitarbeiter erhalten von ihren Arbeiten zusammen 25 Freiexemplare.

Manuscripts should be addressed to/Envoyer les manuscrits à/Manuskripte sind zu senden an:

Prof. Dr. A. BRODAL, Universitetet i Oslo, Anatomisk Institutt, Karl Johans Gate 47 (Domus Media), Oslo 1/Norwegen

Prof. W. HILD, Department of Anatomy. The University of Texas Medical Branch, Galveston, Texas 77550 (USA)

Prof. Dr. J. van LIMBORGH, Universiteit van Amsterdam, Anatomisch-Embryologisch Laboratorium, Amsterdam-O/Holland, Mauritskade 61

Prof. Dr. R. ORTMANN, Anatomisches Institut der Universität, D-5000 Köln-Lindenthal, Lindenburg

Prof. Dr. T. H. SCHIEBLER, Anatomisches Institut der Universität, Koellikerstraße 6, D-8700 Würzburg

Prof. Dr. G. TÖNDURY, Direktion der Anatomie, Gloriastraße 19, CH-8006 Zürich

Prof. Dr. E. WOLFF, Collège de France, Laboratoire d'Embryologie Expérimentale, 49 bis Avenue de la belle Gabrielle, Nogent-sur-Marne 94/France